# Big Data in Medical
# Image Processing

# Big Data in Medical Image Processing

**R. Suganya**
Department of Information Technology
Thiagarajar College of Engineering
Madurai, Tamilnadu, India

**S. Rajaram**
Department of ECE
Thiagarajar College of Engineering
Madurai, Tamilnadu, India

**A. Sheik Abdullah**
Department of Information Technology
Thiagarajar College of Engineering
Madurai, Tamilnadu, India

CRC Press
Taylor & Francis Group
Boca Raton London New York

CRC Press is an imprint of the
Taylor & Francis Group, an **informa** business

A SCIENCE PUBLISHERS BOOK

CRC Press
Taylor & Francis Group
6000 Broken Sound Parkway NW, Suite 300
Boca Raton, FL 33487-2742

First issued in paperback 2020

ISBN-13: 978-1-138-55724-6 (hbk)
ISBN-13: 978-0-367-78151-4 (pbk)

This book contains information obtained from authentic and highly regarded sources. Reasonable efforts have been made to publish reliable data and information, but the author and publisher cannot assume responsibility for the validity of all materials or the consequences of their use. The authors and publishers have attempted to trace the copyright holders of all material reproduced in this publication and apologize to copyright holders if permission to publish in this form has not been obtained. If any copyright material has not been acknowledged please write and let us know so we may rectify in any future reprint.

---

**Library of Congress Cataloging-in-Publication Data**

---

Names: Suganya, R., author.
Title: Big data in medical image processing / R. Suganya, Department of
    Information Technology, Thiagarajar College of Engineering, Madurai,
    Tamilnadu, India, S. Rajaram, Department of ECE, Thiagarajar College of
    Engineering, Madurai, Tamilnadu, India, A. Sheik Abdullah, Department of
    IT, Thiagarajar College of Engineering, Madurai, Tamilnadu, India.
Description: Boca Raton, FL : CRC Press, [2018] | "A science publishers
    book." | Includes bibliographical references and index.
Identifiers: LCCN 2017046819 | ISBN 9781138557246 (hardback : alk. paper)
Subjects: LCSH: Diagnostic imaging--Data processing. | Big data.
Classification: LCC RC78.7.D53 S85 2018 | DDC 616.07/540285--dc23
LC record available at https://lccn.loc.gov/2017046819

---

# Preface

This book covers the syllabus of the various courses like B.E./B.Tech (Computer Science and Engineering, Information Technology, Biomedical Engineering, Electronics and Communication Engineering), MCA, M.Tech (Computer Science and Engineering, Bio Medical Engineering), and other courses related to department of medicine offered by various Universities and Institutions. This book contains the importance of medical imaging in modern health care community.

The chapters involved in this book provide solution for better diagnostic capabilities. The book provides an automated system that could retrieve images based on user's interest to a point of providing decision support. It will help medical analysts to take an informed decisions before planning treatment and surgery. It will also be useful to researchers who are working in problems involved in medical imaging.

The brief contents of this book chapter-wise are given below:

**Chapter 1:** Provides the importance and challenges of Big Data in Medical Image Processing through Hadoop & Map reduce technique.

**Chapter 2:** Starts with Image Pre-processing, importance of speckle in medical images, Different types of filter and methodologies. This chapter presents how to remove speckle noise present in the low modality medical images. Finally this chapter ends with discussion about metrics used for speckle reduction.

**Chapter 3:** Contains the importance of medical image registration, mono modal registration, multi modal image registration. This chapter also covers the procedure involved in the image registration. This chapter deals with optimization techniques like various similarity measures-correlation coefficients and mutual information. Finally this chapter ends with applications of medical image registration with corresponding sample case study.

**Chapter 4:** This chapter begins with introduction on texture analysis and importance of dimensionality reduction. This chapter discusses different types of feature extraction for different medical imaging modalities.

**Chapter 5:** This chapter includes an introduction on machine learning techniques and importance of supervised and unsupervised medical image classification. This chapter discusses various machine learning algorithms like Relevance feedback classifier, Binary vs. multiple SVM, Neural network, Fuzzy classifier with detailed algorithmic representation and simple pictorial representation. Finally this chapter concluded with image retrieval and case study.

## Features

This book has very simple and practical approach to make the readers understand well. It provides how to capture big data medical images from acquisition devices and doing analysis over it. Discuss an impact of speckle (noise) present in the medical images, monitoring the various stages of diseases like cancer and tumor by doing medical image registration. It explains the impact of dimensionality reduction. Finally it acts a recommender system for medical college students for Classifying various stages involved in the diseases by using Machine learning techniques.

# Contents

# 1

# Big Data in Medical Image Processing

## 1.1 An Introduction to Big Data

Big data technologies are being increasingly used for biomedical and healthcare informatics research. Large amounts of biological and clinical data have been generated and collected at an exceptional speed and scale. Recent years have witnessed an escalating volume of medical image data, and observations are being gathered and accumulated. New technologies have made the acquisition of hundreds of terabytes/petabytes of data possible, which are being made available to the medical and scientific community. For example, the new generation of sequencing technologies enables the dispensation of billions of DNA sequence data per day, and the application of electronic health records (EHRs) is documenting large amounts of patient data. Handling out these large datasets and processing them is a challenging task. Together with the new medical opportunities arising, new image and data processing algorithms are required for functioning with, and learning from, large scale medical datasets. This book aims to scrutinize recent progress in the medical imaging field, together with new opportunity stemming from increased medical data availability, as well as the specific challenges involved in Big data. "Big Data" is a key word in medical and healthcare sector for patient care. NASA researchers coined the term big data in 1967 to describe the huge amount of information being generated by supercomputers. It has evolved to include all data streaming from various sources—cell phones, mobile devices, satellites, Google, Amazon, Twitter, etc. The impact of big data is deep, and it will have in-depth implications for medical imaging as healthcare tracks, handles, exploits and documents relevant patient information.

Medical Data collection can necessitate an incredible amount of time and effort, however, once collected the information can be utilized in several ways:

- To improve early detection, diagnosis, and treatment
- To predict patient diagnosis; aggregated data are used to speck early warning symptoms and mobilize resources to proactively address care
- To increase interoperability and interconnectivity of healthcare (i.e., health information exchanges)
- To enhance patient care via mobile health, telemedicine, and self-tracking or home devices

Storing and managing patient health information is a challenging task yet big data in the medical field is crucial. Ensuring patient data privacy and security is also a significant challenge for any healthcare organization seeking to comply with the new HIPAA omnibus rule. Any individual or organization that uses protected health information (PHI) must conform, and this includes employees, physicians, vendors or other business associates, and other covered entities.

HIPAA compliance for data (small or big) must cover the following systems, processes, and policies:

- Registration systems
- Patient portals
- Patient financial systems
- Electronic medical records
- E-prescribing
- Business associate and vendor contracts
- Audits
- Notice of privacy practice

## 1.2 Big Data in Biomedical Domain

In the biomedical informatics domain, big data is a new paradigm and an ecosystem that transforms case-based studies to large-scale, data-driven research. The healthcare sector historically has generated huge amounts of data, driven by record keeping, compliance and regulatory requirements, and patient care. While most data is stored in hard copy form, the current trend is toward rapid digitization of these large amounts of data. Driven by mandatory requirements and the potential to improve the quality of healthcare delivery while reducing the costs, these massive quantities of data (called 'big data') securely hold a wide range of supporting medical and healthcare functions, including amongst others clinical decision support systems, disease outbreak surveillance, and population health management. A disease may occur in greater numbers than expected in a community or region or during a season, while an outbreak may occur in one community or even extend to several countries. July 10, 2017, Measles kills 35 people in Europe as disease spreads through un-vaccinated children, communities are warned by the World Health Organization (WHO). An epidemic occurs when an infectious disease spreads rapidly through a population. For example, in 2003, the severe acute respiratory syndrome (SARS) epidemic took the lives of nearly 800 people worldwide. In Apr 2017, Zika virus is transmitted to people through the bite of an infected mosquito from the *Aedes genus*. This is the same mosquito that transmits dengue, chikungunya and yellow fever. A pandemic is a global disease outbreak. For example, HIV/AIDS is an example of one of the most destructive global pandemics in history.

Reports say data from the U.S. healthcare system alone reached, in 2011, 150 exabytes. At this rate of growth, big data for U.S. healthcare will soon reach the zettabyte (1021 gigabytes) scale and, not long after, the yottabyte (1024 gigabytes). Kaiser Permanente, the California-based health network, which has more than 9 million members, is believed to have between 26.5 and 44 petabytes of potentially rich data from EHRs, including images and annotation.

On 15 May 2017, the Ministry of Health and Family Welfare, Government of India (MoHFW) reported three laboratory-confirmed cases of Zika virus disease in Bapunagar area, Ahmedabad District, Gujarat. National Guidelines and Action Plan on Zika virus disease have been shared with the States to prevent an outbreak of Zika virus disease and containment of spread in case of any outbreak. All the international airports and ports have displayed information for travellers on Zika virus disease. The National Centre for Disease Control, and the National Vector Borne Disease Control Programme are monitoring appropriate vector control measures in airport premises. The Integrated Disease Surveillance Programme (IDSP) is tracking for clustering of acute febrile illness in the community. The Indian Council of Medical Research (ICMR) has tested 34,233 human samples and 12,647 mosquito samples for the presence of Zika virus. Among those, close to 500 mosquitos samples were collected from Bapunagar area, Ahmedabad District, in Gujarat, and were found negative for Zika.

With Zika virus spreading so rapidly, it is really necessary to control the problem on a broader scale. Thus, to put in use analytics-derived insights in this filed, there's a need for ideal platforms that can analyze multiple data sources and create analysis reports. This well-filtered, detailed and real-time information will help vaccine and drug developers to make more powerful vaccines and drugs to fight the disease. The main challenge is how data mining algorithms can track diseases like the Zika virus and help create a different type of response to global disease outbreaks. Based on the outcome of the analysis, neighborhood environment and community health sectors, clinical centers and hospitals can alert patients by presenting with a list of diseases, including zika virus fever, dengue, or information on cases of malaria. The information can also be used to spread more awareness among the public against various diseases.

To fight against Zika, Big Data and analytics can be the major role players as they were in dealing with epidemics such Ebola, flu, and dengue fever. Big Data has already done wonders while dealing with certain complicated global issues and holds broader potential to continue doing so. From a technological perspective, the big data technology can be smartly leveraged to gain insights in how to develop the remedial vaccines for Zika virus by isolating and identifying every single aspect of the virus' characteristics. Although, the statistical modeling and massive data sets are being used across the healthcare community to respond towards the emergency, several big data analytics are still needed to predict these types of contagious diseases. Moreover, the use of technology must be encouraged among the people as well as among the healthcare systems and groups to spread more awareness against the threats, consequences and possible solutions.

## 1.3  Importance of 4Vs in Medical Image Processing

The potential of big data in healthcare lies in combining traditional data with new forms of data, both individually and on a population level. We are already seeing that data sets from a multitude of sources support faster and more reliable research and discovery. If, for example, pharmaceutical developers could integrate population clinical data sets with genomics data, this development could facilitate those developers gaining approvals on more and better drug therapies more quickly than in the past *and*, more importantly, expedite distribution to the right patients. The prospects for all areas of healthcare are infinite. The characteristics of big data is defined by four major Vs such as Volume, Variety, Velocity and Veracity.

### *1.3.1 Volume*

Big data implies enormous volumes of data. First and most significantly, the volume of data is growing exponentially in the biomedical informatics fields. For example, the Proteomics DB covers 92% (18,097 of 19,629) of known human genes that are annotated in the Swiss-Prot database. Proteomics DB has a data volume of 5.17 TB. This data used to be created/ produced by human interaction, however, now that data is generated by machines, networks and human interaction on systems like social media the volume of data to be analyzed is massive. There are several acquisition devices are available to capture medical image modalities. They vary in size and cost. Depending up on the machine, they can capture a huge volume of medical data from human beings. The structured data in EMRs and EHRs include familiar input record fields such as patient name, date of birth, address, physician's name, hospital name and address, treatment reimbursement codes, and other information easily coded into and handled by automated databases. The need to field-code data at the point of care for electronic handling is a major barrier to acceptance of EMRs by physicians and nurses, who lose the natural language ease of entry and understanding that handwritten notes provide. On the other hand, most providers agree that an easy way to reduce prescription errors is to use digital entries rather than handwritten scripts.

Data quality issues are of acute concern in healthcare for two reasons: life or death decisions depend on having the accurate information, and the quality of healthcare data, especially unstructured data, is highly variable and all too often incorrect. (Inaccurate "translations" of poor handwriting on prescriptions are perhaps the most infamous example.)

In the clinical realm, the promotion of the HITECH Act has nearly tripled the adoption rate of electronic health records (EHRs) in hospitals to 44% from 2009 to 2012. Data from millions of patients have already been collected and stored in an electronic format, and this accumulated data could potentially enhance health-care services and increase research opportunities. In addition, medical imaging (e.g., MRI, CT scans) produces vast amounts of data with even more complex features and broader dimensions. One such example is the Visible Human Project, which has archived 39 GB of female datasets. These and other datasets will provide future opportunities for large aggregate collection and analysis.

### 1.3.2 Variety

The second predominant feature of big data is the variety of data types and structures. The ecosystem of biomedical big data comprises many different levels of data sources to create a rich array of data for researchers. Much of the data that is unstructured (e.g., notes from EHRs clinical trial results, medical images, and medical sensors) provide many opportunities and a unique challenge to formulate new investigations. Variety refers to the many sources and different types of data both structured and unstructured. In clinical informatics, there are different variety of data like pharmacy data, clinical data, ECG data, Scan images, anthropometric data and imaging data. We used to store this unstructured data into a clinical repository supported by NoSQL databases. Now medical data comes in the form of emails, photos, videos, monitoring devices, PDFs, audio, etc. This variety of unstructured data creates problems for storage, mining and analyzing data.

### 1.3.3 Velocity

The third important characteristic of big data, velocity, refers to producing and processing data. Big Data Velocity deals with the pace at which data flows in from sources like medical acquisition devices and human interaction with things like social media sites, mobile devices, etc. The speed of the data generated by each radiology centre is deemed to be high. The flow of data is massive and continuous. This real-time data can help researchers and businesses make valuable decisions that provide strategic competitive advantages and ROI if one is able to handle the velocity. The new generation of sequencing technologies enables the production of billions of DNA sequence data each day at a relatively low cost. Because faster speeds are required for gene sequencing, big data technologies will be tailored to match the speed of producing data, as is required to process them. Similarly, in the public health field, big data technologies will provide biomedical researchers with time-saving tools for discovering new patterns among population groups using social media data.

### 1.3.4 Veracity

Big Data Veracity refers to the biases, noise and abnormality in the medical data. In the medical decision support system, veracity plays a vital role in taking decisions about particular diseases or predicting further treatment. The data that is being stored in a clinical decision support system, needs to be mined meaningfully with regards to the problem, and analyzed by several machine learning algorithms. Inderpal (Inderpal et al. 2013) feels veracity in data analysis is the biggest challenge when compared to factors like volume and velocity. Therefore much care is needed to develop any clinical decision support system.

## 1.4 Big Data Challenges in Healthcare Informatics

Big data applications present new opportunities to discover new knowledge and create novel methods to improve the quality of healthcare. The application of big data in healthcare is a fast-growing field. Big data application is a part of four major biomedical sub–disciplines: Bioinformatics, Clinical informatics, Imaging informatics, and Public health informatics. With the advent of widely available electronic health information and Big Data, the massive amount of data produced each day, also provides new opportunities to understand social interactions, environmental and social determinants of health and the impact of those environments on individuals. Big data technologies are increasingly used for biomedical and health-care informatics research. Large amounts of biological and clinical data have been generated and collected at an unprecedented speed and scale.

Informatics is the discipline focused on acquisition, storage and use of information in a specific domain. It is used to analyze the data, manage knowledge, data acquisition and representation, and to manage change and integrate information. Health informatics is classified into six divisions: Clinical Informatics, Medical Informatics, Bioinformatics, Nursing Informatics, Dental Informatics, Veterinary Informatics and Public Health Informatics. Public Health Informatics is an interconnection of healthcare, computer science, and information science. Public Health Informatics (PHI) is defined as the systematic application of information, computer science and technology in areas of public health, including surveillance, prevention, preparedness, and health promotion. The main applications of PHI are:

- promoting the health of the whole population, which will ultimately promote the health of individuals and
- preventing diseases and injuries by changing the conditions that increases the risk of the population.

Primarily, PHI is using informatics in public health data collection, analysis and actions. It deals with the resources, devices and methods required to optimize the acquisition, storage, retrieval and use of information in health and biomedicine. Imaging informatics is the study of methods for generating, managing, and representing imaging information in various biomedical applications. It is concerned with how medical images are exchanged and analyzed throughout complex health-care systems. Imaging informatics developed almost simultaneously with the advent of EHRs and the emergence of clinical informatics.

Specifically, in bioinformatics, high-throughput experiments facilitate the research of new genome-wide association studies of diseases, and with clinical informatics, the clinical field benefits from the vast quantity of collected patient data for making intelligent decisions. Imaging informatics is now more rapidly integrated with cloud platforms to share medical image data and workflows, and public health informatics leverages big data techniques for predicting and monitoring infectious disease outbreaks, such as Ebola. Public health agencies monitor the health status of populations, collecting and analyzing data on morbidity, mortality and the predictors of health status, such as socioeconomic status and educational level. There is a particular focus on the diseases of public health importance, the needs of vulnerable populations and health disparities.

Electronic health information systems can reshape the practice of public health including public health surveillance, disease and injury investigation and control, as well as decision making, quality assurance, and policy development. Scant funding is available to public health departments to develop the necessary information infrastructure and workforce capacity to capitalize on EHRs, personal health records, or Big Data. Types of health information technology (HIT) may play an important role in support of public health including:

- Electronic Health Records (EHRs)
- Personal Health Records (PHR)
- Health Information Exchange (HIE)
- Clinical Decision Support (CDS)

An EHR provides both episodic snapshots and a longitudinal view of a patient's health related to clinical care. A PHR provides a powerful tool for gathering information about clinical visits, since both providers and patients use the application. Many public health departments are pursuing a health-in-all-policies approach to assure that health is a consideration in all major policy decisions. These might include developing new housing, factories, transportation systems, recreation facilities, or educational initiatives to increase graduation rates. Health impact assessments play a critical role in informing decision makers about how their decisions can be used to maximize health and mitigate harm, manage and share health information.

Public health authorities are required to drill down for individual data and risk factors in order to diagnose, investigate and control disease and health hazards in the community, including diseases that originate from social-, environmental-, occupational- and communicable-disease exposures. For example, a clinician or laboratory reports a case of active tuberculosis to the local health department. In response, public health staff performs chart reviews and patient interviews to identify exposed community members and immediately ensure appropriate precautions. For the next year they ensure that all affected patients receive appropriate care and case management. Some of the challenges involved in Big Data in healthcare domains are:

1. Gathering knowledge from complex heterogeneous patient sources
2. Understanding unstructured medical reports in the correct semantic context
3. Managing large volumes of medical imaging data and extracting useful information from it
4. Analyzing genomic data is a computationally intensive task
5. Capturing the patient's behavioural data through several wireless network sensors

## 1.5 Basis of Image Processing

Image processing is a method to perform certain operations on an image, in order to get an enhanced image or to extract some useful information from it. It is a type of signal processing in which the input is an image and the output may be an image or characteristics/features associated with that image. Nowadays, image processing is amongst other rapidly growing technologies. It forms a core research area within engineering and computer science disciplines.

Image processing basically includes the following three steps:

- importing the image via image acquisition tools
- analysing and manipulating the image
- output in which result can be altered image or report that is based on image analysis

There are two types of methods used for image processing namely, analogue and digital image processing. Analogue image processing can be used for the hard copies such as printouts and photographs. Image analysts use various fundamentals of interpretation while using these visual techniques. Digital image processing techniques help in manipulation of the digital images by using computers. The three general phases that all types of data have to undergo while using digital techniques are pre-processing, enhancement and display, and information extraction.

In order to become suitable for digital processing, an image function f(x,y) must be digitized both spatially and in amplitude. Typically, a frame grabber or digitizer is used to sample and quantize the analogue video signal. Hence in order to create an image which is digital, we need to convert continuous data into digital form. There are two steps in which it is done:

- Sampling
- Quantization

The sampling rate determines the spatial resolution of the digitized image, while the quantization level determines the number of grey levels in the digitized image. A magnitude of the sampled image is expressed as a digital value in image processing. The transition between continuous values of the image function and its digital equivalent is called quantization. The number of quantization levels should be high enough for human perception of fine shading details in the image. The occurrence of false contours is the largest problem in images which have been quantized with insufficient brightness levels.

### 1.5.1 Resizing the Image

Image interpolation occurs when one resizes or distorts the image from one pixel grid to another. Image resizing is necessary when one needs to increase or decrease the total number of pixels, whereas remapping can occur when one is correcting for lens distortion or rotating an image. Zooming refers to the process to increase the quantity of pixels, so that when one zooms into an image, one will see more detail.

Interpolation works by using known data to estimate values at unknown points. Image interpolation works in two directions, and tries to achieve a best approximation of a pixel's intensity based on the values at surrounding pixels. Common interpolation algorithms can be grouped into two categories: adaptive and non-adaptive. Adaptive methods change depending on what they are interpolating, whereas non-adaptive methods treat all pixels equally. Non-adaptive algorithms include: nearest neighbour, bilinear, bicubic, spline, sinc, lanczos and others. Adaptive algorithms include many proprietary algorithms in licensed software such as: Qimage, PhotoZoom Pro and Genuine Fractals.

Many compact digital cameras can perform both an optical and a digital zoom. A camera performs an optical zoom by moving the zoom lens so that it increases the magnification of light. However, a digital zoom degrades quality by simply interpolating the image. Even though the photo with digital zoom contains the same number of pixels, the detail is clearly far less than with optical zoom.

### 1.5.2  Aliasing and Image Enhancement

Digital sampling of any signal, whether sound, in digital photographs, or other, can result in apparent signals at frequencies well below anything present in the original. Aliasing occurs when a signal is sampled at a less than twice the highest frequency present in the signal. Signals at frequencies above half the sampling rate must be filtered out to avoid the creation of signals at frequencies not present in the original sound. Thus digital sound recording equipment contains low-pass filters that remove any signals above half the sampling frequency.

Since a sampler is a linear system, then if an input is a sum of sinusoids, the output will be a sum of sampled sinusoids. This suggests that if the input contains no frequencies above the Nyquist frequency, then it will be possible to reconstruct each of the sinusoidal components from the samples. This is an intuitive statement of the Nyquist-Shannon sampling theorem.

Anti-aliasing is a process which attempts to minimize the appearance of aliased diagonal edges. Anti-aliasing gives the appearance of smoother edges and higher resolution. It works by taking into account how much an ideal edge overlaps adjacent pixels.

## 1.6  Medical Imaging

Medical imaging is the visualization of body parts, tissues, or organs, for use in clinical diagnosis, treatment and disease monitoring. Whereas, Medical image processing deals with the development of problem-specific approaches to the enhancement of raw medical image data for the purpose of selective visualization as well as further analysis. There are various needs of medical image processing:

- Hospitals and Radiology center managing several tera-bytes of medical images
- Medical images are highly complex to handle
- Nature of the diseases can be diagnosed by providing solutions that are close to intelligence of doctors

A tactical plan for big data in medical imaging, it is to dynamically integrate medical images, *in vitro* diagnostic information, genetic information, electronic health records and clinical notes into a patient's profile. This provides the ability for personalized decision support by the analysis of data from large numbers of patients with similar conditions. Big data has potential to be a valuable tool, but implementation can pose a challenge in building a medical report with context-specific and target group-specific information that requires access and analysis of big data. The report can be created with the help of semantic technology, an umbrella term used to describe natural language processing, data mining, artificial intelligence, tagging and searching by concept instead of by key word. Radiology can add value to the era of big data by supporting implementation of structured reports.

### 1.6.1 Modalities of Medical Images

Rapid development in the field of the medical and healthcare sector is focused on the diagnosis, prevention and treatment of illness directly related to every citizen's quality of life. Medical imaging is a key tool in clinical practice, where generalized analysis methods such as image pre-processing, feature extraction, segmentation, registration and classification are applied. A large number of diverse radiological and pathological images in digital format are generated by hospitals and medical centers with sophisticated image acquisition devices. Anatomical imaging techniques such as Ultrasound (US), Computed Tomography (CT) and Magnetic Resonance Imaging (MRI) are used daily over the world for non-invasive human examinations.

All the above imaging techniques are of extreme importance in several domains such as computer-aided diagnosis, pathology follow-up, planning of treatment and therapy modification. The information extracted from images may include functional descriptions, geometric models of anatomical structures, and diagnostic assessment. Different solutions such as Picture Archive and Communication Systems (PACS) and specialized systems for image databases address the problem of archiving those medical image collections. The obtained classification results can serve further for several clinical applications such as growth monitoring of diseases and therapy. The main contribution of this research is to address the accuracy of ultrasound liver image classification and retrieval by machine learning algorithms. Among all medical imaging modalities, ultrasound imaging still remains one of the most popular techniques due to its non-ionizing and low cost characteristics. The Digital Imaging and Communication in Medicine (DICOM) Standard is used globally to store, exchange, and transmit medical images. The DICOM Standard incorporates protocols for imaging techniques such as X-ray radiography, ultrasonography, computed tomography (CT), magnetic resonance imaging (MRI) and radiation therapy.

*X-RAY*

X-ray technology is the oldest and most commonly used form of medical imaging. X-rays use ionizing radiation to produce images of a person's internal structure by sending X-ray beams through the body, which are absorbed in different amounts depending on the density of the material. In addition, other devices included as "x-ray type" include mammography, interventional radiology, computed radiography, digital radiography and computed tomography (CT). Radiation Therapy is a type of device which also utilizes x-rays, gamma rays, electron beams or protons to treat cancer. X-ray images are typically used to evaluate:

- Broken bones
- Cavities
- Swallowed objects
- Lungs
- Blood vessels
- Breast (mammography)

*Ultrasound*

Diagnostic ultrasound, also known as medical sonography or ultrasonography, uses high frequency sound waves to create images of the inside of the body. The ultrasound machine sends sound waves into the body and is able to convert the returning sound echoes into a picture. Ultrasound technology can also produce audible sounds of blood flow, allowing medical professionals to use both sounds and visuals to assess a patient's health. Ultrasound is often used to evaluate:

- Pregnancy
- Abnormalities in the heart and blood vessels
- Organs in the pelvis and abdomen
- Symptoms of pain, swelling and infection

*Computer Tomography (CT)*

Computed Tomography (CT), also generally known to as a CAT scan, is a medical imaging method that joins multiple X-ray projections taken from different angles to make detailed cross-sectional images of areas inside the human body. CT images permit doctors to get very precise, 3-D views of certain parts of the body, such as soft tissues, the pelvis, blood vessels, the lungs, the brain, the heart, abdomen and bones. CT is also often the preferred method of diagnosing many cancers, such as liver, lung and pancreatic cancers. CT is often used to evaluate:

- Presence, size and location of tumors
- Organs in the pelvis, chest and abdomen
- Colon health (CT colongraphy)
- Vascular condition/blood flow
- Pulmonary embolism (CT angiography)
- Abdominal aortic aneurysms (CT angiography)
- Bone injuries
- Cardiac tissue
- Traumatic injuries
- Cardiovascular disease

*Magnetic Resonance Imaging (MRI)*

Magnetic Resonance Imaging (MRI) is a medical imaging technology that uses radio waves and a magnetic field to generate detailed images of organs and tissues. MRI has proven to be highly effective in diagnosing a number of conditions by showing the difference between the normal and diseased soft tissues of the body. MRI is often used to evaluate:

- Blood vessels
- Abnormal tissue
- Breasts
- Bones and joints
- Organs in the pelvis, chest and abdomen (heart, liver, kidney, spleen)
- Spinal injuries
- Tendon and ligament tears

*Positron Emission Tomography (PET)*

Positron Emission Tomography (PET) is a nuclear imaging technique that provides physicians with information about how tissues and organs are functioning. PET, often used in combination with CT imaging, and uses a scanner along with a small amount of radiopharmaceuticals which are injected into a patient's vein to assist in making detailed, computerized pictures of areas inside the body. PET is often used to evaluate:

- Neurological diseases such as Alzheimer's and Multiple Sclerosis
- Cancer
- Effectiveness of treatments
- Heart conditions

## 1.7 Medical Image Processing

Image processing is a method to convert an image into digital form and perform certain operations on it, in order to get an enhanced image or to extract some useful information from it. It is a type of signal dispensation in which the input is an image, such as a video frame or photograph and output may be an image or characteristics associated with that image. Usually Image Processing systems includes treating images as two dimensional signals while applying already set signal processing methods to them. It is amongst the rapidly growing technologies today, with its applications in various aspects of a business. Image Processing forms a core research area within the engineering and computer science disciplines too. Image processing basically includes the following three steps.

1. Importing the image with an optical scanner or by digital photography.
2. Analyzing and manipulating the image which includes data compression and image enhancement and spotting patterns that are not visible to human eyes such as satellite photographs.
3. Output is the last stage in which a result can be an altered image or report that is based on image analysis.

Medical Image Processing consists of image preprocessing, image registration, feature extraction, image classification and retrieval. The amount of capturing and analyzing of the biomedical data set is expected to reduce dramatically with the help of technology advances, such as the appearance of new medical machines, the development of new hardware and software for parallel computing, and the extensive expansion of EHRs. Big data applications present new chances to make new discoveries and create novel methods to improve the quality of healthcare. The application of big data in healthcare is a very fast-growing field. The system analyzes medical images then combines this insight with information from the patient's medical records to offer clinicians and radiologists support for decision-making. By applying advanced reasoning and visual technologies, big data analytics in medical imaging filters out the most relevant images that point out abnormalities and provides insight into medical findings.

Today, radiologists may have to monitor as many as 200 cases a day, with some studies containing as many as 3500 images. Each patient's imaging studies could be around 250 GB of data making an institution's image collection petabytes in size. This is where decision support systems are used to considerably reduce diagnosis time and clinician fatigue. The need for Big Data analytics to help sift through all this medical data is clear. Medical images are an important source of data and frequently used for diagnosis, therapy assessment and planning. Computed tomography (CT), magnetic resonance imaging (MRI), X-ray, molecular imaging, ultrasound, photoacoustic imaging, fluoroscopy, positron emission tomography-computed tomography (PET-CT), and mammography are some of the examples of imaging techniques that are well established within clinical settings. Medical image data can range anywhere from a few megabytes for a single study (e.g., histology images) to hundreds of megabytes per study (e.g., thin-slice CT studies comprising upto 2500+ scans per study). Such data requires large storage capacities if stored long term. It also demands fast and accurate algorithms should any decision assisting automation be performed using the data. In addition, if other sources of data acquired for each patient are also utilized during the diagnoses, prognosis, and treatment processes, then the problem of providing cohesive storage and developing efficient methods capable of encapsulating the broad range of data becomes a challenge. The purpose of image processing is divided into 5 groups. They are:

1. **Visualization**—Observe the different modalities of medical regions that are not visible.
2. **Image sharpening and restoration**—To create a better image for diagnosing diseases.
3. **Image retrieval**—Seek for the region of interest by the physicians for future treatment or surgical planning.
4. **Measurement of pattern**—Measures various regions in the image for veracity.
5. **Image recognition**—Distinguish the diseased/infected region in an image.

There are generally five phases in medical image processing. These are Image preprocessing, Image registration or segmentation, Feature extraction, Image classification and Retrieval. All of these five phases together to construct any clinical decision support system shown in Figure 1.

The clinical decision support system takes the input from e-health records and stores it in the clinical repository with anthropometric data for all the patients' case history. Clinical informatics is the study of information technology and how it can be applied to the healthcare field. It includes the study and practice of an information-based approach to healthcare delivery in which data must be structured in a certain way to be effectively retrieved and used in a report or evaluation. If any input from any modalities likes ultrasound, CT scan, MRI. The input image will be converted into a feature vector which is in the matrix format and stored into the database. Feature vector conversion is done in order to reduce the dimensionality reduction. Since medical images are termed as a big data which is associated with 4 Vs called Velocity, Veracity, Volume and Variety. The five major challenges involved in medical image processing are speckle noise, computation time, feature dimensionality, retrieval accuracy and semantic gap.

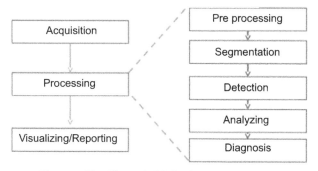

**Figure 1.** Five Phases in Medical Image Processing.

## 1.8 Types of Ultrasound Liver Diseases

### 1.8.1 Liver Cirrhosis

Cirrhosis is a condition in which the liver slowly deteriorates and malfunctions due to chronic injury. Scar tissue replaces healthy liver tissue, partially blocking the flow of blood through the liver. Scarring also impairs the liver's ability to:

- control infections
- remove bacteria and toxins from the blood
- process nutrients, hormones, and drugs
- make proteins that regulate blood clotting
- produce bile to help absorb fats including cholesterol and fat soluble vitamins

A healthy liver is able to regenerate most of its own cells when they become damaged. With end-stage cirrhosis, the liver can no longer effectively replace damaged cells. A healthy liver is necessary for survival.

### 1.8.2 Liver Cyst

A cyst is the medical term used to describe a space of roundish or saclike shape in some part of the body. It may be empty or contain watery or mucus types of fluid. It is not uncommon to find one or several small cysts in the liver when a patient has an ultrasound scan or CAT scan of the abdomen for some reason. The vast majority of these cysts are found by chance as they do not produce any symptoms. It is important to rule out Hydrated disease. This condition occurs when humans ingest dog tape worm, Echinococcus granulose which can invade the liver causing cysts. This may occur in areas where sheep and cattle raising are done. These can be differentiated from simple cysts, this is important as they may require treatment to avoid rupturing. Good standards of hygiene and regular deworming of dogs can prevent these infections.

### 1.8.3 Normal Liver

This is the external surface of a normal liver. The color is brown and the surface is smooth. A normal liver is about 1200 to 1600 grams. The liver is the largest organ in the body. It is located on the right side of the abdomen (to the right of the stomach) behind the lower ribs and below the lungs. The liver performs more than 400 functions each day to keep the body healthy. Some of its major jobs include:

- converting food into nutrients the body can use (for example, the liver produces bile to help break down fats)
- storing fats, sugars, iron, and vitamins for later use by the body
- making the proteins needed for normal blood clotting
- removing or chemically changing drugs, alcohol, and other substances that may be harmful or toxic to the body

### 1.8.4 Fatty Liver

Fatty liver, also known as fatty liver disease (FLD) is a reversible condition where large vacuoles of triglyceride fat accumulate in liver cells via the process of steatosis (i.e., abnormal retention of lipids within a cell). Despite having multiple causes, fatty liver can be considered a single disease that occurs worldwide in those with excessive alcohol intake and those who are obese (with or without effects of insulin resistance). The condition is also associated with other diseases that influence fat metabolism. Morphologically it is difficult to distinguish alcoholic FLD from non-alcoholic FLD and both show micro-vesicular and macrovesicular fatty changes at different stages.

Accumulation of fat may also be accompanied by a progressive inflammation of the liver (hepatitis), called steatohepatitis. By considering the contribution by alcohol, fatty liver may be termed alcoholic steatosis or non-alcoholic fatty liver disease (NAFLD), and the more severe forms as alcoholic steatohepatitis (part of alcoholic liver disease) and non-alcoholic steatohepatitis (NASH).

## 1.9 Challenges in Medical Images

### 1.9.1 Image Pre-processing—Speckle Reduction

A major problem for handling the medical images is the presence of various granular structures such as speckles noise. Sometimes the speckle contains fine details of the image content, especially related to diagnostic features used by humans in unassisted diagnosis. Speckle is the random granular texture that obscures anatomy in ultrasound images usually called noise. Speckle is created by a complex interference of ultrasound echoes made by reflectors spaced closer together than the machine's resolution limits. The issue of speckle can be reduced with higher frequency imaging, but this of course would limit the depth of the ultrasound penetration. Speckle cannot be directly correlated with reflectors or cells and are thus an artifact of ultrasound technology.

SRI, Speckle Reduction Imaging is the first real-time algorithm that removes speckle without the disadvantages that have plagued other methods. The adaptive nature of the SRI algorithm allows it to smooth regions where no feature or edges exist, while maintaining and enhancing edges and borders. It has been shown that SRI increases contrast resolution by increasing signal to noise ratio. Additionally, the algorithm does not eliminate any information so diagnostic criteria is preserved. Such image quality improvements can increase consistency in diagnosis, reduce patient and operator dependence and ultimately improve diagnostic competency and confidence.

Ultrasound Image Reflector Location Envelope Data Magnification of optical and ultrasound images illustrate that speckle seen in an ultrasound image is not directly related to physical structure such as a liver cell

Amplitude and Phase Sum Resolved Reflectors Resolution Limit Complex Interference 20 micron.

SRI, Speckle Reduction Imaging, is an advanced image processing technique to remove speckle and is available exclusively on ultrasound systems. SRI is based on the most recent advances in real-time adaptive image processing and is enabled by the computational power of Ultrasound Architecture. The SRI algorithm works by analyzing the image pixel-by-pixel, and classifying each as 'Mostly Speckle' or 'Mostly Feature'. This is done by comparing neighboring pixels to see if variations have a sharp difference, follow a trend or are random in nature. Random variations are considered random in nature and are thus suppressed. By accurately recognizing the difference between speckle and feature, SRI is able to appropriately suppress speckle while maintaining feature resulting in an image that looks more like the actual tissue and provides greater diagnostic confidence. SRI algorithm on a typical liver showing: (a) edge enhancement (b) speckle suppression (c) feature presentation Original SRI Table showing the dramatic decrease in variation and increase in SNR with using SRI By dramatically improving signal to noise ratio, the benefits from SRI are similar to those from crossbeam although the method of image processing is entirely different.

### 1.9.2 Image Registration

Another important issue in medical image classification and retrieval is the computational time that occurs during registration of diseases. Image registration plays a key role in medical image analysis procedure. The computational time of the image registration algorithms greatly varies depending on the type of the registration computed, and the size of the image to process. Image registration is the process of combining two or more images for providing more information. Medical image fusion refers to the fusion of medical images obtained from different modalities. Medical image fusion helps in medical diagnosis by way of improving the quality of the images. In diagnosis, images obtained from a single modality like MRI, CT, etc. may not be able to provide all the required information. There is a need to combine information obtained from other modalities to improve the information acquired. For example a combination of information from an MRI and CT modalities provides more information than the individual modalities separately.

Image registration is a primary step in many real time image processing applications. Registration of images is the bringing of two or more images into a single coordinate system for its subsequent analysis. It is sometimes called image alignment. It is widely used in remote sensing, medical imaging, target recognition using multi-sensor fusion, monitoring of usage of a particular land using satellite images, images alignment obtained from different medical modalities for diagnosis of diseases. It is an important step in the field of image fusion and image mosaicing.

The image registration methods can be grouped into two classes. One is intensity based method which is based on gray values of the pair of images and the second one is based on image feature which is done by obtaining some features or landmarks in the images like points, lines or surfaces. Edges in the images can be detected very easily in the images. Thus, using these edges some features can be obtained by which we can accomplish feature based registration. But, feature based registration has some limitations as well as advantages. The proposed method employs feature based registration techniques to obtain a coarsely registered image which can be given as input to intensity based registration techniques to get a fine registration result. It helps to reduce the limitations of intensity based technique, i.e., it takes less time for registration. To achieve this task, the mutual information is selected as similarity parameter.

Mutual information (MI) is used widely as a similarity measure for registration. In order to improve the robustness of this similarity measure, spatial information is combined with normalized mutual information (NMI). MI is multiplied with a gradient term to integrate spatial information to mutual information and this is taken as similarity measure. The registration function is less affected if sampling resolution is low. It contains correct global maxima which are sometimes not found in case of mutual information. For an optimization purpose, Nealder mead, Fast Convergence Particle Swarm Optimization technique (FCPSO) is generally suggested. In this optimization method, the diversity of position of single particle is balanced by adding a new variable, particle mean dimension (pmd) of all particles to the existing position and velocity equation. It reduces the convergence time by reducing the number of iterations for optimization. There are two types of medical image registration: Mono-modal and Multi-modal. Mono modal means registration is done on the two same modality of images (for example: US-US, MRI-MRI). Multi-modal image registration deals with two images with different modalities. For example—MRI-CT.

Intensity-based automatic image registration is an iterative process. It requires that one specify a pair of images, a metric, an optimizer, and a transformation type. In this case the pair of images is the MRI image (called the reference or fixed image) which is of size 512*512 and the CT image (called the moving or target image) which is of size 256*256. The metric is used to define the image similarity metric for evaluating the accuracy of the registration. This image similarity metric takes the two images with all the intensity values and returns a scalar value that describes how similar the images are. The optimizer then defines the methodology for minimizing or maximizing the similarity metric. The transformation type that is used is rigid transformation (2-Dimension) that works translation and rotation for a target image (brings the misaligned target image into alignment image with the reference image). Before the registration process can begin the two images (CT and MRI) need to be preprocessed to get the best alignment results. After the preprocessing phase the images are ready for alignment. The first step in the registration process was specifying the transform type with an internally determined transformation matrix. Together, they determine the specific image transformation that is applied to the moving image. Next, the metric compares the transformed moving image to the fixed image and a metric value is computed. Finally, the optimizer checks for a stop condition. In this case, the stop condition is the specified maximum number of iterations. If there is no stop condition, the optimizer adjusts the transformation matrix to begin the next iteration. And display the results of it in part of the result.

### 1.9.3 Image Fusion and Feature Extraction

In clinical diagnosis, the amount of texture in the pathology bearing region (PBR) is one of the key factors used to assess the progression of liver diseases. Feature set of high dimensionality causes the "curse of dimension" problem, in which the complexity of computational cost of the query increases exponentially with the number of dimensions. The next stage after the registration process is the wavelet based image fusion. Wavelets are finite duration oscillatory functions with a zero average value. They can be described by two functions the scaling (also known as the father) function, and the wavelet (also known as the mother) function. A number of basic functions can be used as the mother wavelet for Wavelet Transformations. The mother wavelet through translation and scaling produces various wavelet families which are used in the transformations.

The wavelets are chosen based on their shape and their ability to analyze the signal in a particular application. The Discrete Wavelet Transform has the property that the spatial resolution is small in low-frequency bands but large in high-frequency bands. This is because the scaling function is treated as a low pass filter and the wavelet function as a high pass filter in DWT implementation. The wavelet transform decomposes the image into low-high, high-low, high-high spatial frequency bands at different scales

22

and the low-low band at the coarsest scale. The low-low image has the smallest spatial resolution and represents the approximation information of the original image. The other sub images show the detailed information of the original image.

### 1.9.4 Image Segmentation

Image segmentation is a procedure for extracting the region of interest (ROI) through an automatic or semi-automatic process. Many image segmentation methods have been used in medical applications to segment tissues and body organs. Some of the applications consist of border detection in angiograms of coronary, surgical planning, simulation of surgeries, tumor detection and segmentation, brain development study, functional mapping, blood cells automated classification, mass detection in mammograms, image registration, heart segmentation and analysis of cardiac images, etc.

In medical research, segmentation can be used in separating different tissues from each other, through extracting and classifying features. One of the efforts is classifying image pixels into anatomical regions which Medical Image Segmentation Methods, Algorithms, and Applications, may be useful in extracting bones, muscles, and blood vessels. For example, the aim of some brain research, works is to partition the image into different region colours such as white and different grey spectra which can be useful in identifying the cerebrospinal fluid in brain images, white matter, and grey matter (Medina et al. 2012). This process can also prove useful in extracting the specific structure of breast tumors from MRI image.

A region is composed of some pixels which two by two are neighbours and the boundary is made from differences between two regions. Most of the image segmentation methods are based on region and boundary properties. Here we explain two most popular region-based approaches: thresholding and region growing.

### 1.9.5 Image Classification and Retrieval

One of the main challenges for medical image classification and retrieval is to achieve meaningful mappings between the high-level semantic concepts and the low-level visual features (color, texture, shape) called semantic gap. Another main challenge in classification and retrieval of medical image is retrieval accuracy. Sometimes medical images with fat deposits might be unpredicted as images with cysts, because of the similar appearance in case of manual diagnosis. Hence to overcome these kinds of misdiagnosis, it is essential to develop a system that would classify the images into proper categories. It is also very important to facilitate a proper retrieval system that would serve as a tool of guidance for the physician in treatment.

## 1.10  Introduction to Data Classification

Data mining algorithms are classified into three different learning approaches: supervised, unsupervised and semi-supervised. In supervised learning, the algorithms works with a set of examples whose labels are well-known. The labels can be small values (nominal) in the case of the classification task, or numerical values in the case of the regression task. In unsupervised learning, quite the reverse, the labels of the examples in the dataset are unidentified, and the algorithm usually aims at grouping examples according to the resemblance of their attribute values, characterizing a clustering job. As a last step semi-supervised learning is usually used when a small subset of labelled examples is offered, together with a massive number of unlabeled examples.

The classification task can be seen as a supervised method where every instance fit in to a class, which is specified by the value of a unique goal attribute or basically the class attribute. The goal attribute can take on definite values, each of them corresponding to a class. Each example consists of two parts, specifically a set of predictor attribute values and a goal attribute value. The former are used to predict the value at the last point. The predictor attributes should be significant for predicting the class of an instance. In the classification task the set of examples being extracted is divided into two mutually exclusive and exhaustive sets, called the training set and the test set. The classification process is respectively divided into two phases:

- **Training Phase:** In this phase, a classification model is built from the training set.
- **Testing Phase:** In this phase, the model is evaluated on the test set.

In the training phase, the algorithm has the right to use to the values of both predicator attributes and the goal attributes for all illustrations of the training set, and it utilizes that information to construct a classification model. This model represents classification information—basically, a relationship between predictor attribute values and classes—that permit the forecast of the class of an illustration given its predictor attribute values. For the testing phase, the test set, class values of the examples are not exposed. In the testing phase, only once a prediction is made is the algorithm authorized to see the actual class of the just classified example. One of the key goals of a classification algorithm is to exploit the predictive accuracy obtained by the classification model when classifying examples in the test set unseen throughout the training phase.

In some cases, such as lazy knowledge, the training phase is absent entirely, and the classification is performed straight from the relationship of the training instances to the test example. The output of a classification algorithm may be presented for a test instance in one of the two ways:

1. **Discrete tag:** In this case, a tag is returned for the test instance.
2. **Numbers set:** In this case, a numbers set can be converted to a discrete label for a test instance, by selecting the class with the elevated set for that test instance.

Advancements in imaging analytics are making algorithms increasingly capable of doing interpretations currently done by radiologists. Algorithms can analyze the pixel and other bits and bytes of data contained within the image to detect the distinct patterns associated with a characteristic pathology. The outcome of the algorithmic analysis is a metric. In the current early stage of imaging analytics, these metrics complement the analysis of the images made by radiologists, and help them render a more accurate or a faster diagnosis. For example, it is possible today to calculate bone density by applying an algorithm on any CT image of a bone. The resulting number is then compared with a threshold metric to determine whether the patient is at risk of fracture. If the number is below the threshold, a doctor can prescribe a regular intake of calcium or other preventative measure. The screening for 'low bone density' is made automatically without a dedicated and additional exam. It is determined simply by leveraging an existing CT examination performed on a patient. This is an important first step into preventative care.

The development of these automated analysis tools is already under way. Research teams and start-up companies across the world work every day to produce new algorithms to cover more body parts and pathologies. It won't take long before radiologists are equipped with thousands of predictive algorithms to automatically detect the patterns of the most standard diseases. This application of advanced data analysis holds the exciting prospect of preventing diseases.

## 1.11 Data Classification Technologies

In this segment, the different methods that are frequently used for data classification will be discussed. The most common methods used in data classification are decision trees, Support Vector Machine methods, Naive

Bayesian method, instance-based method and neural networks. The classification among technologies is illustrated in Figure 2.

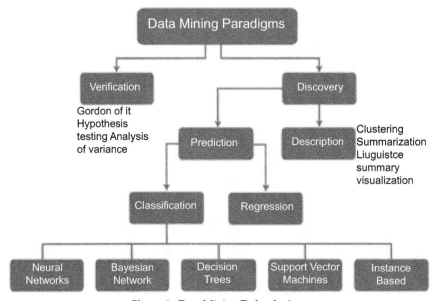

**Figure 2.** Data Mining Technologies.

## 1.12 Data Mining Technique

In this section, technical aspects of data mining techniques are used to analyze WSN dataset are described (Gupta et al. 2005). Wireless Sensor Networks are networks that consist of sensor motes which are scattered in an ad hoc manner. These motes work with each other to sense certain physical facts and then the information gathered is processed to get appropriate results. Wireless sensor networks consist of protocols and algorithms with self-organizing capabilities. Motes are a physical unit in Wireless sensors used to explore novel data mining techniques, and dealing with extracting knowledge from large communities deriving data from WSNs. Some of the data mining techniques for WSNs datasets are described in the Table 1.

As the term mining refers to the extraction of mineral resources from earth. Data Mining is a process that works on the numerous data that is

**Table 1.** Data Mining Techniques for Medical Data.

| Data Mining tech | Algorithms | Handling Medical Data |
|---|---|---|
| Frequent and Sequential pattern mining | Apriori and Growth-based algorithms | To find association among large low modality medical data set |
| Cluster based technique | K-means, hierarchical and data correlation based clustering | Based upon the distance among the data point |
| Classification based technique | Decision tree, Random forest, NN, SVM and Logistic regression | Based on the application—Medical applications |

available in the database, in order to extract the useful information. Data Mining has been widely used in various fields because of its efficiency in pattern recognition, and knowledge discovery in databases.

Evolution of data mining was observed from 1960s. Initially people were focused on collection of data which specified various techniques for gathering data such as surveys, interviews, online transactions, customer reviews, etc. This is the initial task to be performed in any data mining. After which the Data warehousing techniques, which use the online Analytic processing for the decision support is used. Data Mining uses the advanced methodology such as predictive analytics, machine learning, etc. for its purpose.

It has been used in various domains such as the medical field, marketing, social media, mobile service providers, etc. In the marketing field it can be used to predict the products that need promotion, sets of products that are purchased together, for targeted marketing, etc. In social media it can be used for sentiment analysis, recent trend analysis, etc. Mobile service providers use this technique to promote a particular package based on the usage of services, calculating the talk-time for every user and using this as a database to perform analytics, which can be used to create a new service packs, etc. In the medical field it can be used to predict diseases based on the symptoms provided, and combination of medicine to be used for curing diseases, etc.

Data Mining can be classified into Text Mining, Image Mining, Video Mining, and Audio Mining. Data mining has been carried out in the following order. First the data is collected from the respective sources which are the most relevant for our application, then the preprocessing

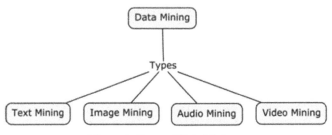

**Figure 3.** Types of Data Mining.

step will be carried out, this deals with the removal of unwanted data, conversion of data into the form that is useful for processing. Using data mining techniques such as prediction, description, etc. In this study we have taken Image mining as our area of interest. Image mining has been used for weather forecasting, investigation in police departments, for locating the enemy camp in the military field, finding minerals, the medical field, etc.

Let us consider the medical domain where there has been huge advancement in recent years due to the tremendous increase in use of internet. This leads to the e-healthcare system, to send patients information such as patient name, identification number, age, location, previous medical reports, diagnosis reports, etc. through the internet. This is leading to the new era of digital images sent to various hospitals to be used for future diagnosis. While sending this information, we should consider the fact that the information can be modified, tampered, lost, falsified, etc. during transmission. So care should be taken to prevent these kinds of attacks during transmission.

### 1.12.1 Decision Tree

Decision trees are recursive partitioning algorithms used to minimize the impurity present in the sensor dataset. The apex node is the root node specifying a testing provision of which the outcome corresponds to a branch leading up to an internal node. The fatal nodes of the tree allot the classifications and are also referred to as the leaf nodes. Popular decision trees are C4.5 CART and Chi-Squared Analysis. This algorithm consists of splitting decision, stopping decision and assignment decision. The tree will start to fit the specificities or noise in the data, which is referred to as overfitting. In order to avoid this, the sensor data will be split into a training sample and validation sample. The training example will be used to construct the splitting assessment. The validation sample is an independent sample used to supervise the misclassification error. One way to determine impurity in a node is by calculating the mean squared error (MSE).

Decision tree is a classifier in the form of a tree and classifies the occurrence by starting at the root of the tree and moving through it until a leaf node where class label is assigned. The internal nodes are used to partition data into subsets by applying test conditions to separate instances that have different characteristics. Decision tree learning is a process for resembling discrete-valued target functions, in which the learned function is represented by a decision tree. It generates a hierarchical partitioning of the data, which relates the different partitions at the leaf level to different classes. Decision trees classify illustrations by sorting them down the tree from the root to a few leaf nodes, which provide the classification of the instances. Each node in the tree specifies a test of some attributes of the illustration, and each branch falling from that node corresponds to one of the possible values for this attribute. An occurrence is classified by starting at the root node of the tree, testing the attribute specified by this node, then shifting down the tree branch corresponding to the value of the attribute in the given example. This procedure is then repeated for the sub tree rooted at the new node (Breiman et al. 1992).

The hierarchical partitioning at each level is created with the use of an inference split criterion. The inference split criterion may either use a condition on a single element or it may contain a condition on multiple elements. The former is referred to as a univariate split, while the last is referred to as a multivariate split. The objective is to pick the attribute that is the most useful for classifying examples. The overall approach is to try to recursively split the training data so as to maximize the bias among the diverse classes over diverse nodes. The discrimination among the different classes is maximized, when the point of skew among the diverse classes in a given node is maximized. A measure such as gini-index or entropy is used in order to quantify this skew.

For example if q1......qk is the fraction of the records belonging to the k different classes in a node N, then the gini-index G(N) of the node N is defined as follows:

$$G(N) = 1 - \sum_{i-1}^{k} qi^2$$

The value of G(N) lies between 0 and 1-1/k. The lesser the significance of G(N), the superior the skew. In these cases where the classes are regularly balanced, the value is 1-1/k. The alternative measure is entropy E(N):

$$E(N) = -\sum_{i-1}^{k} qi \cdot \log(qi)$$

The value of the entropy lies between 0 and log(k). The value is log(k), when the records are perfectly balanced among the different classes. This matches to the scenario with maximum entropy. The smaller the entropy is, the greater the skew in the data. Thus gini-index and entropy provide an effectual way to assess the quality of a node in terms of its level of discrimination between the different classes.

Algorithm Decision Trees

begin

01: create a node N

02: for d=1 to number of training observations and its class values

03:         for a=1 to number of candidate attributes

04:                 Select a splitting criterion

05:             end for

06: end for

07: Create a node $N_d$

08: if all observations in the training dataset have the same class output value C, then

09:     return $N_d$ as a leaf node labeled with C

10:     if attribute list = {∅}, then

11:             return $N_d$ as a leaf node labeled with majority class output value.

12:             Apply selected splitting criterion

13:             Label node $N_d$ with the splitting criterion attribute

14:             Remove the splitting criterion attribute from the attribute list

15:     for each value i in the splitting criterion attribute

16:             $D_i$ = no. of observations in training dataset satisfying attribute value i

17:     if $D_i$ is empty then

18:             attach a leaf node labeled with majority class output value to node $N_d$

19: else

20:             attach the node returned by decision tree to node $N_d$

21:     end if

22: end for

23:             return node $N_d$

24:     end if

25: end if

There are various specific decision-tree algorithms.

- ID3 (Iterative Dichotomiser 3)
- C4.5 (Successor of ID3)
- CART (Classification and Regression Tree)
- CHAID (Chi-Squared Automatic Interaction Detector)
- MARS: extends decision trees to handle numerical data better

### 1.12.2 Support Vector Machines (SVM)

SVM was first introduced by Vapnik and has been a very effective method for regression, classification and pattern recognition. It is measured as a good classifier because of its high generalization recital without the necessitate to add apriori facts, even when the measurements of the input space is very high. The goal of SVM is to find the finest classification function to distinguish between parts of the two classes in the training data. SVM methods employ linear circumstances in order to split out the classes from one another. The design uses a linear condition that separates the two classes from each other as well as possible. Consider the medical application, where the risk of ultrasound liver diseases are related to diagnostic features from patients. SVM is used as a binary classifier to predict whether the patient is diagnosed with a liver disease or not. In such a case, the split condition in the multivariate case may also be used as stand-alone condition for classification. This, SVM classifier, may be considered a single level decision tree with a very carefully chosen multivariate split condition. The effectiveness of the approach depends only on a single separating hyper plane. This separation is challenging to define.

Support Vector Machine is a supervised learning technique used for classification purposes. For supervised learning, a set of training data and category labels are available and the classifier is designed by exploiting this prior known information. The binary SVM classifier uses a set of input data and predicts each given input, classifying where the two possible classes of data belongs. The original data in a finite dimensional space is mapped into a higher dimension space to make the separation easier. The vector classified closer to the hyper plane is called support vectors. The distance between the support vector and the hyper plane is called the margin; the higher marginal value given the lower the error of the classifier. The separation among higher and lower dimensions is described in Figure 4.

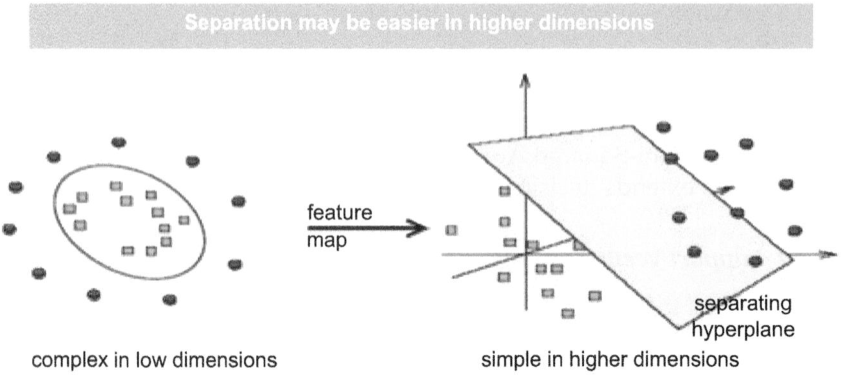

Figure 4. Dimensions for Support Vector Machines.

Support Vector Machines are normally labelled for binary classification problems. Therefore, the class variable yi for the ith training instance Xi is assumed to be drawn from {–1, +1}. The most significant criterion, which is commonly used for SVM classification, is the maximum margin hyper plane. The metric for the concept of the "best" classification function can be recognized geometrically. For a linearly separable dataset, a linear classification role matches to a separating hyper plane f(X) that bypasses through the centre of the two classes, separating the two.

Once this function is determined, new data instance f(x, n) can be classified by simply testing the sign of the function f(xn); Xn fit in to the positive class if f(Xn) > 0. Since there are a lot of such linear hyper planes, SVM assures that the best such function is established by maximizing the margin between the two classes. This separation among the classes is described in Figure 5.

Naturally, the margin is stated as the amount of space, or partition between the two sets of data as defined by the hyper plane. Geometrically, the margin matches to the shortest distance between the closest data point to a point on the hyper plane. This is in order to ensure that the maximum margin hyper planes are actually established.

One of the initial disadvantages of SVM is its computational ineptitude; however this dilemma is being solved with immense success. One approach to solve the computational ineptitude is to split a large optimization dilemma into a series of smaller dilemmas, where each dilemma only engages with a couple of carefully chosen variables, so that the optimization can be ended powerfully. This process repeats itself until all the decomposed

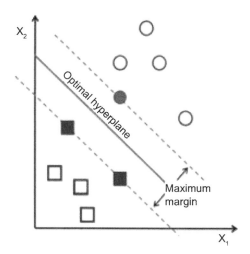

**Figure 5.** Marginal values in a Hyper Plane.

optimization problems are solved successfully. Another newer approach used for solving the SVM optimization problem is Sequential Minimal Optimization methods; this method considers the problem of learning an SVM, as that of finding an approximate minimum set of instances (Jiawei et al. 2011).

### *1.12.3 Neural Network*

A Neural Network or Simulated Neural Network (SNN) is an interconnected group of artificial neurons that use a statistical or computational model for information processing based on a connection approach to computation. In several cases an ANN is an adaptive system that changes its structure based on external or internal information which flows through the network. NN are non-linear statistical data modeling or decision making tools. They can be used to model multifarious relationships between inputs and outputs or to find similar patterns in data (Pluim et al. 2003). The procedure adopted for NN is as follows: (a) Train a neural network for WSNs data as much as possible in terms of connection, Split the data into training, validation and test set. (b) Assign the number of hidden neurons as 5 and categorize the hidden unit activation values using clustering. (c) Extract rules that explain the network output and inputs in terms of the categorized hidden unit activation rules and train a neural network on the training set with anthropometric data and measure the performance on the validation set (d) Choose the number of hidden neurons with optimal validation set performance. (e) Measure the performance on the independent test set.

A Neural Network or Simulated Neural Network (SNN) is an interconnected cluster of artificial neurons that employ a statistical or computational model for information processing based on a correlation approach to computation. Neural networks were modelled after the cognitive processes of the brain. In the human brain, neurons are connected to one another via points, which are referred to as synapses. They are capable of predicting new observations from existing observations. This biological analogy is retained in an artificial neural network. The basic computation unit in an ANN consists of unified processing elements also called units, nodes, or neurons. These neurons can be arranged in different kinds of architecture by the connections between them. The neurons within the network work together, in parallel, to produce an output task. Since the computation is achieved by the collective neurons, a neural network can still produce the output function if a few number of the individual neurons are malfunctioning (the network is robust and fault tolerant).

In general, each neuron inside a neural network has a related activation number. In addition, each connection between neurons has a weight correlated with it. These quantities replicate their counterparts in the biological brain: firing rate of a neuron, and power of a synapse. The activation of a neuron depends on the activation of the other neurons and the weight of the edges that are related to it. The neurons within a neural network are regularly arranged in layers. The quantity of layers within the neural network, and the quantity of neurons within each layer normally matches the nature of the examined fact.

After the size has been decided, the network is usually then focused to training. Here, the network accepts a sample training input with its related classes. It then applies an iterative process on the input in order to fine-tune the weights of the network so that its future forecasts are best possible. After the training phase, the network is ready to perform forecasts in new sets of data (Hastie et al. 2001).

Neural networks can frequently construct very accurate predictions. However, one of their greatest criticisms is the fact that they represent a "black-box" approach to investigation. Neural networks do not offer any insight into the underlying nature of the phenomena. The most basic architecture of the neural network is a perceptron, which holds a set of input nodes and an output node. The output unit receives a set of inputs from the input units. There are different input units which are exactly equal to the dimensionality of the underlying data. The data is supposed to be numerical. Categorical data may need to be transformed into binary representations, and therefore the number of inputs may be larger. The output node is associated with a set of weights W, which are used in order to compute a function f(.) of its inputs. Each component of the weight vector is associated with a connection from the input unit to the output unit. The weights can be viewed as the analogue of the synaptic strengths in biological systems. In case of perceptron architecture, the input nodes do not carry out any computations. They simply pass the input attribute forward. Computations are performed only at the output nodes in the basic perceptron architecture. The output node uses its weight vector along with the input attribute values in order to compute a function of the inputs. A typical function, which is computed at the output node, is the signed linear function:

$$Zi = \sin\{W.Xi + b\}$$

The output is a predicted value of the binary class variable, which is assumed to be drawn from {−1, +1}. The notation b denotes the bias. Thus, for a vector Xi drawn from a dimensionality of d, the weight vector W should also contain d elements. Now consider a binary classification problem, in which all labels are drawn from (+1, −1}. We assume that the class label of Xi is denoted by yi. In that case, the sign of the predicted function zi yields the class label. The input layers with activation function are described in Figure 6.

In the case of single layer perceptron algorithms, the training process is easy to carry out by using a gradient descent approach. The major challenge in training multilayer networks is that it is no longer known for intermediate (hidden layer) nodes, what their "expected" output should be. This is only

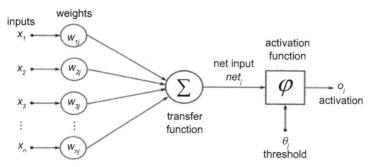

**Figure 6.** Neural Network Model.

known for the final output node. Therefore, some type of "error feedback" is needed, in order to choose the changes in the weights at the intermediate nodes. The training process proceeds in two phases, one of which is in the forward direction, and the other is in the backward direction.

### 1.12.4 Sequential Minimal Optimization

Sequential Minimal Optimization (SMO) is an algorithm for proficiently solving the optimization problem which arises during the training of SVM samples. It was introduced by John Platt in 1998 at Microsoft research. SMO is widely used for training SVM. SVM techniques separate the data belonging to different classes by fitting a hyperplane between them which maximized the partition. The data is mapped into a higher dimensional feature space where it can easily partitioned by a hyperplane. SVM-Linear kernel is adopted for predicting heart diseases using WSNs dataset.

### 1.12.5 Random Forest

Random forest was first introduced by Breiman et al. (1992). It creates a forest of decision trees as follows:

1. Given a data set with 13 observations and N inputs from 303 data instances
2. Assume m = constant
3. For t = 1….T
   a. Take a bootstrap sample with 13 observations
   b. build a decision tree w hereby for each node of the tree, randomly choose m inputs on which to base the splitting decision based on the sensor information from motes
   c. split on the best of this subset
   d. fully grow each tree without pruning

## 1.13 Big Data Technologies

Biomedical scientists are facing new challenges of storing, managing, and analyzing massive amounts of datasets. The characteristics of big data require powerful and novel technologies to extract useful information and enable more broad-based health-care solutions. In most of the cases reported, we found multiple technologies that were used together, such as artificial intelligence (AI), along with Hadoop, and data mining tools.

Parallel computing is one of the fundamental infrastructures used for managing big data tasks. It is capable of executing algorithm tasks simultaneously on a cluster of machines or supercomputers. In recent years, novel parallel computing models, such as MapReduce by Google, have been proposed for new big data infrastructures. More recently, an open-source MapReduce package called Hadoop was released by Apache for distributed data management. The Hadoop Distributed File System (HDFS) supports concurrent data access to clustered machines. Hadoop-based services can also be viewed as cloud-computing platforms, which allow for centralized data storage as well as remote access across the Internet.

As such, cloud computing is a novel model for sharing configurable computational resources over the network and can serve as an infrastructure, platform, and/or software for providing an integrated solution. Furthermore, cloud computing can improve system speed, agility, and flexibility because it reduces the need to maintain hardware or software capacities and requires fewer resources for system maintenance, such as installation, configuration, and testing. Many new big data applications are based on cloud technologies.

## 1.14  Hadoop and MapReduce Technique

Advances in information and communication technology present the most viable solutions to big data analysis in terms of efficiency and scalability in medical arena. It is vital that those big data solutions are multithreaded and that data access approaches be precisely tailored to large volumes of semi-structured/unstructured data. Hadoop is an open source software implementation of the MapReduce framework for running applications on large clusters built of commodity hardware from Apache. Hadoop is a platform that provides both distributed storage and computational capabilities. Hadoop was first comprehended to fix a scalability issue that existed in Nutch, an open source crawler and search engine that utilizes the MapReduce and big-table methods developed by Google. Hadoop is a distributed master-slave architecture that consists of the Hadoop Distributed File System (HDFS) for storage and the MapReduce programming framework for computational capabilities. The HDFS stores data on the computing nodes providing a very high aggregate bandwidth across the cluster.

Traits inherent to Hadoop are data partitioning and parallel computation of large datasets. Its storage and computational capabilities scale with the addition of computing nodes to a Hadoop cluster, and can reach volume sizes in the petabytes on clusters with thousands of nodes. Hadoop also provides Hive and Pig Latin, which are high-level languages that generate MapReduce programs. Several vendors offer open source and commercially supported Hadoop distributions; examples include Cloudera, DataStax, Hortonworks and MapR. Many of these vendors have added their own extensions and modifications to the Hadoop open source platform.

Hadoop differs from other distributed system schemes in its philosophy toward data. A traditional distributed system requires repeat transmissions of data between clients and servers. This works fine for computationally intensive work, but for data-intensive processing, the size of the data becomes too large to be moved around easily. Hadoop focuses on moving code to data instead of vice versa. The client (NameNode) sends only the MapReduce programs to be executed, and these programs are usually small (often in kilobytes). More importantly, the move-code-to-data philosophy applies within the Hadoop cluster itself. Data is broken up and distributed across the cluster, and as much as possible, computation on a chunk of data takes place on the same machine where that chunk of data resides.

The MapReduce programming framework uses two tasks common in functional programming: Map and Reduce. MapReduce is a new parallel processing framework and Hadoop is its open-source implementation on a single computing node or on clusters. Compared with existing parallel processing paradigms (e.g., grid computing and graphical processing unit (GPU)), MapReduce and Hadoop have two advantages: (1) fault-tolerant storage resulting in reliable data processing by replicating the computing tasks, and cloning the data chunks on different computing nodes across the computing cluster; (2) high-throughput data processing via a batch processing framework and the Hadoop distributed file system (HDFS). Data is stored in the HDFS and made available to the slave nodes for computation.

In this book, we discussed the existing applications of the MapReduce programming framework and its implementation platform Hadoop in clinical big data and related medical health informatics fields. The usage of MapReduce and Hadoop on a distributed system represents a significant advance in clinical big data processing and utilization, and opens up new opportunities in the emerging era of big data analytics. The objective of this paper is to summarize the state-of-the-art efforts in clinical big data analytics and highlight what might be needed to enhance the outcomes of clinical big data analytics tools. This book is concluded by summarizing the potential usage of the MapReduce programming framework and Hadoop platform to process huge volumes of clinical ultrasound liver data in medical health informatics related fields.

## 1.15 NoSQL Databases for Storing Medical Images

Since medical images are called big data with 4Vs, it is not suitable for Relational Data base Management system. RDBMS works well for structured data and follows ACID property. Whereas huge volumes of different modalities of big data should be handled by an NoSQL database. NoSQL DB is otherwise called Schema less DB. There are four different types of NoSQL databases. They are Key-value pair, Document Database, Graph database and Columnar Database.

### 1.15.1 Key-Value Database

A *key-value database* (also known as a key-value store and key-value store database) is a type of NoSql database that uses a simple key/value method to store data. The key-value part refers to the fact that the database stores data as a collection of key/value pairs. This is a simple method of storing data, and it is known to scale well. The key-value pair is a well established concept in many programming languages. Programming languages typically refer to a key-value as an associative array or data structure. A key-value is also commonly referred to as a dictionary or hash. In key values store, there is no schema and the value of the data is opaque. Values are identified and accessed via a key, and the stored values can be numbers, strings, medical images, videos and more. The key value data store is mainly used to maintain clinical support systems. Many implementations that are not well-suited to traditional RDBMS can benefit from a key-value models which offers several benefits including:

**Flexible Data Modelling:** Because a key-value store does not enforce any structure on the data, it offers tremendous flexibility for modelling data to match the requirements of the application.

**High Performance:** Key-value architecture has higher performance than relational databases in many scenarios because there is no need to perform lock, join, union, or other operations when working with objects. Unlike traditional relational databases, a key-value store doesn't need to search through columns or tables to find an object. Knowing the key will enable very fast location of an object.

**Massive Scalability:** Most key-value databases make it easy to scale out on demand using commodity hardware. They can grow to virtually any scale without significant redesign of the database.

**High Availability:** Key-value databases may make it easier and less complex to provide high availability than that which can be achieved with relational database. Some key-value databases use a masterless, distributed architecture that eliminates single points of failure to maximize resiliency.

**Operational Simplicity:** Some key-value databases are specifically designed to simplify operations by ensuring that it is as easy as possible to add and remove capacity as needed and that any hardware or network failures within the environment do not create downtime.

### 1.15.2 Document Data Store

A document store database (also known as a document-oriented database, aggregate database, or simply document store or document database) is a database that uses a document-oriented model to store data. Document store databases store each record and its associated data within a single document like a single patient history. It includes anthropometric data, clinical data, pharmacy data, scan reports, surgery reports, hospital admission reports, etc. as a single document. Each document contains semi-structured data that can be queried against using various query and analytics tools of the DBMS. Relational databases store data within multiple tables, each table containing columns, and each row represents each record. Information about any given entity could be spread out among many tables. Data from different tables can only be associated by establishing a relationship between the tables.

Document databases on the other hand, don't use tables as such. They store all data on a given entity within a single document. Any associated data is stored inside that one document. With relational databases, one must create a schema before one can load any data. With document store databases (and most other NoSQL databases), one has no such requirement. One can just go ahead and load the data without any predefined schema. So with a document store, any two documents can contain a different structure and data type. For example, if one user chooses not to supply his date of birth, that wouldn't be a field within the document. If another user does supply her date of birth, that would be a field in that document. If this was a relational database, date of birth would still be a field for both users—it just wouldn't contain a value.

Document databases can scale horizontally very well for more than 1000 patients in the hospital. Data can be stored over many thousands of computers and the system will perform well. This is often referred to as sharding. Relational databases are not well suited to scaling in this fashion. Relational DBs are more suited towards scaling vertically (i.e., adding more memory, storage, etc.). Seeing as there's a limit to how many resources one can fit inside one machine, there could come a point where horizontal scaling becomes the only option. Document stores don't have foreign keys, like relational databases have. Foreign keys are used by relational databases to enforce relationships between tables. If a relationship needs to be established with a document database, it would need to be done at the application level.

However, the whole idea behind the document model is that any data associated with a record is stored within the same document. So the need to establish a relationship when using the document model should not be as prevalent as in a relational database. Most relational databases use SQL as the standard query language. Document store databases tend to use other query languages (although some are built to support SQL). Many document databases can be queried using languages such as XQuery, JSON, SPARQL. Document databases are similar to key-value databases in that, there's a key and a value. Data is stored as a value. Its associated key is the unique identifier for that value. The difference is that, in a document database, the value contains structured or semi-structured data. This structured/semi-structured value is referred to as a document. The structured/semi-structured data that makes up the document, can be encoded using one of any number of methods, including XML, JSON, YAML, BSON, etc. It could also be encoded using binary, such as PDFs, MS Office documents, etc.

### A Benefit of the Document Model over Key-Value Stores

One benefit that document store databases have over key-value databases is that one can query the data itself. One can query against the structure of the document, as well as the elements within that structure. Therefore, one can return only those parts of the document that one requires. With a key-value database, one gets the whole value—no matter how big (and seemingly structured) it might be. One can't query within the value.

### 1.15.3 *Mongo Database*

Mongo DB is a cross platform document oriented database used for storing medical dataset. It makes the integration of different modalities of application easier and faster. Developed by the Mongo DB Corporation. Inc. Mongo DB has been adopted as the backed software by a number of major website and services especially for storing medical images, E-health records, Public health records, ECG datasets, anthropometric datasets, etc.

*Features of Mongo DB*

1. *Document Oriented*
   Mongo DB can store the medical dataset in a minimal number of documents. For Example instead of storing Patient name and anthropometric data in two distinct relational structures, Patient name and anthropometric data and other disease related information can be stored in a single document.

2. *Adhoc Queries*
   Mongo DB supports the search by field range queries, regular expression searches are allowed. Search can be done either by medical images for future diagnosis or pharmacy details of a particular patient or by his ECG report, etc. Queries can return a specific field of documents and also include user defined java scripts functions.

3. *Indexing*
   Any field in a Mongo DB document can be indexed similar to RDBMS. Secondary indexes are also available. Meta datas are stored by master file in the HADOOP. Meta data contains information about particular patient details, history of his health, clinical data, medical imaging, etc.

4. *Replication*
   Mongo DB provides high availability with replica datasets. Replica dataset consists of two or more copies of data. The primary copy performs all the read/write operation by default. The secondary copy maintains the copy of primary replicas using built in replicas. This feature is particularly helpful for doctors located in different countries for diagnosing same diseases.

5. *Load Balancing*
   Mongo DB scales horizontally using Sharding. The user chooses a shard key which determines how the medical data in a collection will be distributed.

6. *File Storage*
   Mongo DB can be used as a file system taking advantage of load balancing and data replication features over multiple missions for storing the medical files. This function called Grid FS is included with Mongo DB drivers and available with no difficulty for development

languages. Instead of storing huge medical files in a single document, Grid File System divides the files into parts or chunks and stores each of the chunks in separate document.

### 1.15.4 Columnar Database

A column store database is a type of database that stores data using a column oriented model and is shown in the Figure 7. A column store database can also be referred to as a: Column database, Column family database, Column oriented database, Wide column store database, Wide column store, Columnar database and Columnar store. Column store databases are considered NoSQL databases, as they use a different data model to relational databases.

Columns store databases use a concept called a keyspace. A keyspace is like a schema in the relational model. The keyspace contains all the column families (similar to tables in the relational model), which contain rows, and which contain columns.

For Example:

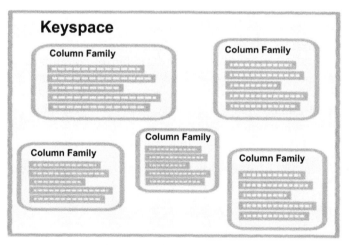

**Figure 7.** Columnar Data Base.

A keyspace containing column families, which include a complete history of his/her health records.

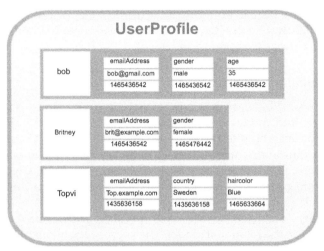

**Figure 8.** Column Family Representation.

A column family containing 3 rows. Each row contains its own set of columns.

As the above diagram shows:

- A **column family** consists of multiple rows.
- Each **row** can contain a different number of columns to the other rows. And the columns don't have to match the columns in the other rows (i.e., they can have different column names, data types, etc.).
- Each **column** is contained to its row. It doesn't span all rows like in a relational database. Each column contains a name/value pair, along with a timestamp. Note that this example uses Unix/Epoch time for the timestamp.

Here is how each row is constructed:

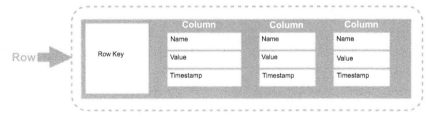

**Figure 9.** Column-Row Family Representation.

Here is a breakdown of each element in the row:

- **Row Key.** Each row has a unique key, which is a unique identifier for that row.
- **Column.** Each column contains a name, a value, and timestamp.
- **Name.** This is the name of the name/value pair.
- **Value.** This is the value of the name/value pair.
- **Timestamp.** This provides the date and time that the data was inserted. This can be used to determine the most recent version of data.

Some DBMSs expand on the column family concept to provide extra functionality/storage ability. For example, Cassandra has the concept of *composite columns*, which allows one to nest objects inside a column.

*Benefits of Column Store Databases*

Some key benefits of columnar databases include:

- **Compression.** Column stores are very efficient at data compression and/or partitioning.
- **Aggregation queries.** Due to their structure, columnar databases perform particularly well with aggregation queries (such as SUM, COUNT, AVG, etc.).
- **Scalability.** Columnar databases are very scalable. They are well suited to massively parallel processing (MPP), which involves having data spread across a large cluster of machines—often thousands of machines.
- **Fast to load and query.** Columnar stores can be loaded extremely fast. A billion row table could be loaded within a few seconds. One can start querying and analysing almost immediately.

These are just some of the benefits that make columnar databases a popular choice for organisations dealing with big data.

# 2

# Image Processing

## 2.1 Introduction

Image processing is any form of signal processing for which the input is an image, such as a photograph or video frame; the output of image processing may be either an image or a set of characteristics or parameters related to the image. Most image-processing techniques involve treating the image as a two-dimensional signal and applying standard signal-processing techniques to it.

Image processing usually refers to digital image processing, but optical and analog image processing are also possible. This chapter is about general techniques that apply to all of them. Medical imaging is the technique and process used to create images of the human body for clinical purposes or medical science. Although imaging of removed organs and tissues can be performed for medical reasons, such procedures are not usually referred to as medical imaging, but rather are a part of pathology.

Texture is one of the important characteristics used in identifying and characterizing objects or regions of interest in an image. The commonly-used texture analysis approaches are based on the probability distribution of gray levels and the texture pattern properties. The features are measured from the first-order and second-order statistics. Examples of the first-order statistical parameters that are employed include the mean, variance, skewness and kurtosis of the one-dimensional gray level histogram. Layer observed a significant increase in the 'mean gray level of the ultrasound B-scan image of liver with a regional arrangement of large fat deposits. However, no difference in image brightness was observed between normal liver tissue and liver steatosis with diffuse homogeneous fatty infiltration. In general, the first-order statistical parameters have been difficult to apply in practice since they depend significantly on the gain or amplification settings of the ultrasound equipment.

The second-order statistical methods include the gray-level co-occurrence matrices (GLCM) and the gray-level run-length matrices. Haralick et al. (1979) proposed a set of 14 features calculated from a co-occurrence matrix whose elements represent estimates of the probability of transitions from one gray level to another in a given direction at a given intermixed distance. The features derived from GLCM include contrast, entropy, angular second moment, sum average, sum variance and measures of correlation. This technique has been shown to discriminate normal from abnormal liver, spleen pathologies, prostate disease, myocardial ischemia, human placenta and beef IMFAT. Parkin et al. (2005) showed that GLCM can be applied on different intermixed distances to reveal periodicity in the texture. However, there is an inherent problem to choose the optimal interpixel distance in a given situation. Also, the GLCM method, in general, is not efficient since a new co-occurrence matrix needs to be calculated for every selected angle and inter-pixel distance. The main difficulty of the above methods is due to the lack of an adequate tool that characterizes different scales of textures effectively. Recent developments in spatial-scale analysis such as Gabor transform, Wigner distribution and wavelet transform have provided a new set of multiresolution analytical tools. The fundamental idea underlying wavelets is the ability to analyze the signal at different scales or resolutions. The wavelet analysis procedure uses a scalable wavelet prototype function, called the mother wavelet. The fine frequency analysis is performed with a contracted, high-frequency version of the wavelet function. The coarse frequency analysis is performed with a dilated, low-frequency version of the same wavelet. This multiscale or multiresolution view of signal analysis is the essence of the wavelet transforms.

Many researchers have studied wavelet-based multi-resolution for image texture classification and segmentation. Chang proposed a tree-structured wavelet transform and applied it to texture classification. This new structure makes it possible to zoom into any desired frequency channel for further decomposition. The authors claim the algorithm outperformed several other conventional methods. Wu et al. used wavelet-decomposed images to derive rotation and gray-scale transform invariant texture features. The five phases of image processing are shown in Figure 10.

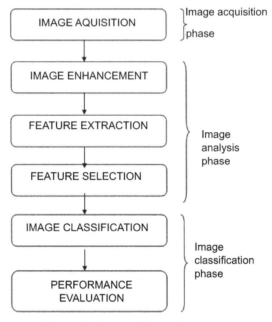

**Figure 10.** Phases of Image Processing.

## 2.2 Speckle Removal

To improve image analysis, speckle reduction is generally used for two applications:

- Visualization enhancement
- Auto-segmentation improvement

Most speckle filters are developed for enhancing visualization of speckle images. Medical imaging like Ultrasound is very popular due to its low cost, least harm to human body, real time view and small size. Filter analysis has been done to select the best filter for a given image on the basis of some statistical parameters. Finally we designed the fused method which has various combinations of filters to give a better outcome. Among all medical imaging modalities, ultrasound imaging still remains one of the most popular techniques due to its low power, easy-to-use nature and low cost characteristics. As ultrasound waves pass through a body, they are reflected back due to constructive and destructive interference of the ultrasound machine in different ways, based on the characteristics of the tissues encountered. One of the major problems of ultrasound images is the presence of noise, having the form of a granular pattern, called speckle. Automatic interpretation of ultrasound images however is extremely difficult because of its low signal to noise ratio (SNR). One of the main reasons for this low SNR is the presence of speckle noise.

It is generally desirable for image brightness (or film density) to be uniform except where it changes to form an image. There are factors, however, that tend to produce variation in the brightness of a displayed image even when no image detail is present. This variation is usually random and has no particular pattern. In many cases, it reduces image quality and is especially significant when the objects being imaged are small and have relatively low contrast. This random variation in image brightness is designated noise. This noise can be either image dependent or image independent. All the digital images contain some visual noise. The presence of noise gives an image a mottled, grainy, textured, or snowy appearance. There are three primary types of noise: Random, fixed pattern and banding. Random noise revolves around an increase in intensity of the picture. It occurs through color discrepancies above and below where the intensity changes. It is random, because even if the same settings are used, the noise occurs randomly throughout the image. It is generally affected by exposure length. Random noise is the hardest to get rid of because we cannot predict where it will occur. Banding noise depends on the camera as not all digital cameras will create it. During the digital processing steps, the digital camera takes the data being produced from the sensor and creates the noise from that. High speeds, shadows and photo brightening will create banding noise. Gaussian noise, salt & pepper noise, passion noise, and speckle noise are some of the examples of noise.

Speckle noises tend to obscure diagnostically important features and degrade the image quality and thus increase the difficulties in diagnosis. The automated recognition of meaningful image components, anatomical structures, and other pathology bearing region, is typically achieved using some kind of speckle reduction techniques.

Though much work has been done for the speckle reduction in ultrasound images in the literature, most of the work assumes that the diffusion based spatial filter will suppress the noise and preserve fine details of edges. But it lacks in diagnosing small residual features like cyst and lesion that will lead to major problems like cirrhosis and cancer that finally lead to death. Much care must be taken to identify small pathology bearing information like cyst and lesion that is hidden by means of speckles during pre-processing.

## 2.3 Importance of Speckles in Medical Images

Diagnostic usage for ultrasound has greatly expanded over the past couple of decades because of the advantages of low cost and non-invasive nature. Ultrasound, however, suffers from an inherent imaging artifact called speckle. Speckle is created by a complex interference of ultrasound echoes made by reflectors. A speckle depends on the ultrasound system's resolution limit. The ability of diagnosing ultrasound images to detect low contrast lesions is fundamentally limited by speckles. Speckle is a granular pattern formed due to constructive and destructive coherent interferences of backscattered signals due to unresolved tissue in-homogeneity. The speckle pattern depends on the structure of the imaged tissue and various imaging parameters, e.g., the frequency and geometry of the ultrasound transducer. Because of its dependence on the microstructure of the tissue parenchyma, speckle is often used in diagnosis such as focal and diffuse liver diseases. Since, noise shows up as speckle, it reduces image contrast and obscures image details. So, handling speckles in ultrasound images is a critical task in medical image processing. After removing speckles present in ultrasound images, depending on the user needs, the image is passed into either Image registration for growth monitoring of pathology changes in liver or Feature extraction for classification & retrieval.

The characteristics of the imaging system based on the spatial distribution of speckles can divide speckles into three classes (Wagner et al. 1983) as given below:

i) The fully formed speckle pattern occurs when much randomly distributed scattering exists within the resolution cell of the imaging system. For example—Blood cells.

ii) The second class of tissue scatters is random distribution with long-range order. Example of this type is lobules in liver parenchyma.

iii) The last class occurs when invariant coherent structure is present within the random scatter region like organ surfaces and blood vessels.

There are two approaches used for speckle reduction.

## 2.4 Types of Filter

The speckle reduction in ultrasound images is a challenging task in medical image processing. Several research works have been reported in the literature for analyzing the speckle reduction in ultrasound images. The first work is the adaptive filtering of speckle introduced in ultrasound pulse-echo images to determine the amount of smoothing according to the ratio of local variance to local mean. In this method, smoothing is increased in uniform texture regions, where speckle is fully developed and reduced or even avoided in other regions to preserve details.

## 2.5 Related Works

Perona and Malik (1990) developed anisotropic diffusion to overcome the problem in standard scale space technique. In this diffusion process, the removal of noise and preservation of edges are simultaneously achieved by choosing the conduction coefficient as a function. This technique can simultaneously eliminate noise and preserve or even enhance edges. Because of this attractive feature, many researchers have applied anisotropic diffusion techniques in speckle noise reduction for medical ultrasound images. This method has supported our work.

Catte et al. (1992) used a smoothed gradient of the image, rather than the true gradient. The smoothing operator removes some of the noise which might have deceived the original PM filter (Perona and Malik 1990). In this case, the scale parameter is fixed. Pyramid transform has also been used for reducing speckle (Aiazzi et al. 1998). Sattar et al. (1997) adopted Feavueau's pyramid transform. In this method, edge detection on interpolated detailed images is used to determine which of the pixels in the detailed images are to be included in the reconstruction stage. However, the existing edge detection approaches are sophisticated enough to perform well on ultrasound images. Due to the multiplicative nature of speckle noise, Aiazzi et al. (1998) introduced a laplacian pyramid transform. In this method, the existing Kuan filter (Kuan et al. 1985) is extended to multiscale domain by processing the interscale layers of the ratio laplacian pyramid, but it suffers noise variance in each interscale layer.

Czerwinski et al. (1995) proposed a directional median filter to the problem of boundary-preserving speckle reduction in ultrasound images. The technique applies a bank of one-dimensional median filters and retains the largest value among all filter's output at each pixel. The result is an operator that suppresses speckle noise while retaining the structure of the image, particularly the thin bright streaks that tend to cause boundaries between tissue layers.

Yu and Acton (2002) introduced an edge sensitive diffusion method [(i.e., speckle reducing anisotropic diffusion (SRAD)] to suppress speckle while preserving edge information. However, the SRAD is only for uncompressed echo envelope images, and its performance declines when it is directly applied to log compressed images. In addition, a speckle detector that simply combines the gradient magnitude and laplacian may not perform well for the boundaries between regions with different gray levels.

Abd-Elmoniem et al. (2002) presented a tensor-based anisotropic diffusion method, i.e., Nonlinear Coherent Diffusion (NCD) for speckle reduction and coherence enhancement. This approach combines three different models: isotopic diffusion, anisotropic coherent diffusion and mean curvature motion. The disadvantage of the NCD model is that a nonselective Gaussian smoothing filter is needed before estimating structure tensors, which may eliminate feature details smaller than the smoothing kernel. Both of the above two diffusion method can preserve or even enhance prominent edges when removing speckles, but they have one common limitation in retaining subtle features, such as small cysts and lesions in ultrasound images.

Chen et al. (2003) proposed Region growing-based spatial filtering methods. In this method, it is assumed that pixels that have similar gray level and connectivity are contextually related and likely to belong to the same object or region. After all pixels are allocated to different groups, spatial filtering is performed based on the local statistics of adaptive regions whose size and shapes are determined by the information content of the image. The main difficulty in applying this method is how to design appropriate similarity criteria for region growing.

Yongjian Yu et al. (2004) derived a Partial Differential Equation (PDE) for speckle reduction from minimizing a cost functional of the instantaneous coefficient of variation. Comparison between SRAD and generalized SRAD shows that the generalized SRAD is complementary to the conventional SRAD, in the sense that the former one emphasizes feature preservation and the latter emphasizes edge enhancement, while both remove speckle in the observed images.

Most of the wavelet thresholding methods suffer from a drawback that the chosen threshold may not match the specific distribution of signal and noise components in different scales. To address this issue, Gupta et al. (2005) presented sub band coefficients of the log-transformed ultrasound image and it can be successfully modeled using the generalized laplacian distribution. Based on this modeling, a simple adaptation of the zero zone and reconstruction levels of the uniform threshold quantizer are proposed in order to achieve simultaneous de-speckling and quantization.

Khalifa Djemal (2005) proposed speckle deduction in ultrasound images by minimization of total variation. To limit the noise in an image, some techniques are based on the calculation of an average intensity in each pixel of the image by considering some neighborhood. However, these techniques tend to attenuate contours present in the image. This affects edge for segmentation and makes difficult for proper decision making.

Fan Zhang et al. (2006) proposed laplacian pyramid nonlinear diffusion and shock filter (LPNDSF). LPNDSF couples nonlinear diffusion and shock filter process and is applied in Laplacian pyramid domain of an image to remove speckle and enhance edges simultaneously. The LPND model proposed by Fan Zhang et al. (2007) incorporates the favorable properties of multiscale analysis in noise-signal separation into anisotropic diffusion. Because the laplacian pyramid domain could achieve improved noise suppression and edge preservation. But LPND works only on log compressed data.

Shin-Min Chao et al. (2010) proposed a diffusion model which incorporates both the local gradient and the gray-level variance to preserve edges and fine details while effectively removing noise. When the level of noise is high; noisy pixels in the image generally involve larger magnitudes of gray level variance and gradients than those of actual edges and fine details. Thus, the method may soon be found to be inefficient. In (2010), Surya Prasath et al. proposed the inhomogeneous anisotropic diffusion which includes a separate multiscale edge detection part to control the diffusion.

Analysis of ultrasound images can assist in differentiating between benign and malignant lesions. It is the imaging modality of choice in the evaluation of suspected adnexal masses (Togashi 2008) as well as liver masses (Kinkel et al. 2000). According to Kinkel et al. (2000), ultrasound is the main triage method for liver disease prior to treatment.

The authors Guodong et al. combined cellular neural network (CNN) and gray step co-occurrence matrix to process B-scan images of patients' fatty livers. They dealt with the B-scan images of patients' fatty livers by the edge detection cellular neural network, and then analyze the B-scan image features, including the co-occurrence matrix's contrast (Contrast), correlation (Correlation), energy (Energy) and homogeneity (Homogeneity). The value of Contrast on 0° direction seems to correlate to the degree of the damage of patients' livers. It is expected that the method provided in this paper will be helpful in the diagnosis of biomedical images.

S. Mukherjee, A. Chakravorty concluded that seven textural feature descriptors are calculated from the SGLD matrices for each of the ultrasound subimages of seventy-six normal livers and twenty-four fatty livers at different pixel pair differences and orientations. A one dimensional Self-organizing map (SOM) is used to identify the representative feature vectors. Two components of the feature vectors, namely maximum probability and uniformity, are identified to be most distinguishing on the basis of their statistical distribution. These two components are used as inputs to a two dimensional SOM grid and after convergence the neurons are plotted in their weight space. Superposition of these plots for normal and fatty liver images clearly shows clusters with little or no overlap between them. They have estimated the degree of distinction between the normal and fatty clusters with the help of quality factor and identified the optimal combination of pixel pair distance and orientation. This is a humble beginning to model the radiologists' perceptual findings that may emerge in future as a new tool with respect to 'ultrasonic biopsy'.

Coocurrence features showed better segmentation performance for 32 X 32 and 16 X 16 pixel window resolutions while histogram features showed better segmentation performance for 8 X 8, 4 X 4 and 2 X 2 pixel window resolutions.

The performance of the maximum likelihood classifier has been tested on natural textures and it is concluded by Metin kahraman.

To address the unsuitability of the above approach for multiplicative speckled situations, this chapter proposes a modified laplacian pyramid nonlinear diffusion for speckle reduction in ultrasound images. Image pre-processing by speckle reduction work is completed by concentrating on improving edge information using Modified Laplacian Pyramid Non-linear Diffusion filter.

## 2.6 Phases of Pre-processing

Pre-processing is the most important phase in medical image analysis. Pre-processing is generally done to prepare the data set for the analysis part. The data set must be of less error so that accurate analysis can be done. The three major steps in pre-processing are the following:

- Resizing
- Speckle noise removal
- Contract adjustment

### 2.6.1 *Resizing*

Scaling is a non-trivial process that involves a trade-off between efficiency, smoothness and sharpness. As the size of an image is increased, so the pixels which comprise the image become increasingly visible, making the image appears "soft". Conversely, reducing an image will tend to enhance its smoothness and apparent sharpness. Apart from fitting a smaller display area, image size is most commonly decreased in order to produce thumbnails. Enlarging an image is generally common for making smaller imagery fit a bigger screen in full screen mode as shown in Figure 11, for example. In "zooming" an image, it is not possible to discover any more information in the image than already exists, and image quality inevitably suffers. However, there are several methods of increasing the number of pixels that an image contains, which evens out the appearance of the original pixels. The data set collected are images. They are resized to the default size of 100 x 100 pixels.

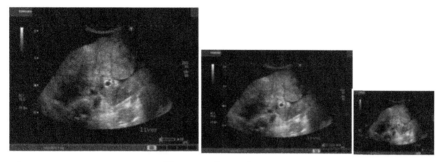

**Figure 11.** Image Resize.

### 2.6.2 Speckle Noise Removal by Anisotropic Diffusion

We are looking for an efficient method of noise removal that is able to clarify the image. Methods of noise removal were known a long time before the Anisotropic Diffusion case was first claimed.

In anisotropic diffusion the main motto is to encourage smoothening within the region in preference to the smoothening across the edges. This is achieved by setting the conduction coefficient as 1 within the region and as 0 near edges, however the main problem involved in this is the detection of the presence and absence of edges. As a solution for this problem it is identified that conduction coefficient, if chosen locally as a function of magnitude of the gradient of the brightness function of the image, the edges can be determined. A general expression for anisotropic diffusion can be written as $I(x, 0) = I_0$

$$\frac{\partial I}{\partial t} = div(F) + \beta (I_0 - I)$$

where, I is the input image, $I_0$ is the initial image, F is the diffusion flux and $\beta$ is a data attachment coefficient. If $\beta = 0$, particular cases of equation are:

55

The non linear probability density function (PDF) with
F = c ( | ΔI | ) .I. where
Δ is the gradient operator,
*div* is the divergence operator,
|| denotes the magnitude, and diffusion coefficient c (x) is given by:

$$C(x) = \frac{1}{1+(x/k)}$$

and

$$C(x) = exp[\frac{1}{1+(x/k)}]$$

the functions for the diffusion coefficient:

$$C(\|\nabla I\|) = e^{-(\|\nabla I\|/K)^2}$$

$$c(\|\nabla I\|) = \frac{1}{1+\left(\dfrac{\|\nabla I\|}{K}\right)^2}$$

and the constant K controls the sensitivity to edges and is usually chosen experimentally or as a function of the noise in the image. The anisotropic diffusion function is shown in Figure 12.

### 2.6.3 Contrast Limited Adaptive Histogram Equalization (CLAHE)

Each time an image is acquired, window and level parameters must be adjusted to maximize contrast and structure visibility. This must be done before the image is saved in any other format than the generic format of the acquisition software HIS. For the moment, very little post-processing in addition to window-level is applied to the image after its acquisition. This is due in part to the good quality of the image without processing, but also because of the short experience and tools we have working with 16 bit images.

CLAHE seems a good algorithm to obtain a good looking image directly from a raw HIS image, without window and level adjustment. This is one possibility to automatically display an image without user intervention. Further investigation of this approach is necessary. CLAHE was originally developed for medical imaging and has proven to be successful for enhancement of low-contrast images such as portal films.

The CLAHE algorithm partitions the images into contextual regions and applies the histogram equalization to each one. This evens out the

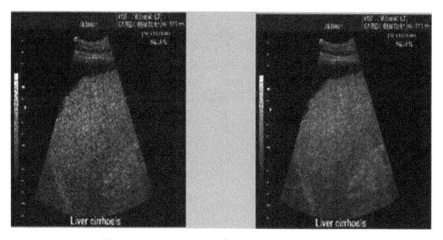

**Figure 12.** Anisotropic Diffusion Applied to Image.

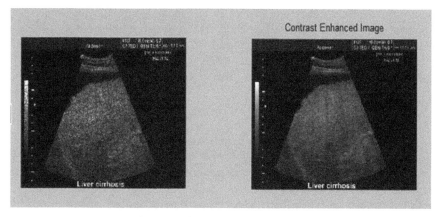

**Figure 13.** Contrast Enhanced Image.

distribution of used grey values and thus makes hidden features of the image more visible. The full grey spectrum is used to express the image. Contrast Limited Adaptive Histogram Equalization, CLAHE, is an improved version of AHE, or Adaptive Histogram Equalization and is shown in Figure 13. Both overcome the limitations of standard histogram equalization. A variety of adaptive contrast-limited histogram equalization techniques (CLAHE) are provided. Sharp field edges can be maintained by selective enhancement within the field boundaries. Selective enhancement is accomplished by first detecting the field edge in a portal image and then only processing those regions of the image that lie inside the field edge.

Noise can be reduced while maintaining the high spatial frequency content of the image by applying a combination of CLAHE, median filtration and edge sharpening. This technique known as Sequential processing can be recorded into a user macro for repeat application at any time. A variation of the contrast limited technique called adaptive histogram clip (AHC) can also be applied. AHC automatically adjusts clipping level and moderates over enhancement of background regions of portal images.

$$I = \frac{(I_{max} - I_{min})}{(I_{max} - I_{min})}$$

**Where**

$I_{max}$ is the healthy region
$I_{min}$ is the infected region

## 2.7 Laplacian Pyramid Domain

The first step in the proposed approach is the transformation of input liver image into its laplacian pyramid domain. The Laplacian pyramid was first introduced by Burt and Adelson in (1983). A structure of laplacian pyramid domain in the proposed MLPND consists of two stages:

  i)  Pyramid decomposition
  ii) Pyramid reconstruction

Both can be described by approximation and interpolation filtering. However, the laplacian pyramid has one advantage compared to wavelet filter bank, i.e., each pyramid level generates only one bandpass image which does not scramble frequencies. In the pyramid decomposition stage, a signal is successively decomposed into a decimated approximation signal and a signal containing small important information. This small signal is computed as the difference between the signals on a scale and the interpolated signal from a rough texture region. A smooth texture region corresponds to a lower pyramid layer whereas rough texture region corresponds to higher pyramid layer. The lowest pyramid layer has equal size as in the original image. A specific pyramid domain is determined by its approximation and interpolation filter plays a major role in finding small spots.

In the Laplacian pyramid, two operators are used for Pyramid decomposition and reconstruction.

i) REDUCE—The REDUCE operator carries out a two-dimensional (2-D) lowpass filtering on ultrasound image followed by a sub-sampling factor of two in both directions.

ii) EXPAND—The EXPAND operator enlarges an ultrasound image to twice the size in both directions by up-sampling factor of four.

For an input ultrasound liver image $I$, let its Gaussian pyramid at layer $l$ be $G_1$, and its Laplacian pyramid at layer $l$ be $L_1$, where $l = 0, 1, 2 \ldots d–1$ and $d$ is the total decomposition layer (d = 5). Then, the Gaussian and Laplacian pyramid can be defined as:

$$G_0 = 1 \tag{1}$$
$$G_1 = REDUCE[G_{1-1}] \tag{2}$$
$$L_1 = G_1 - EXPAND[G_{1+1}] \tag{3}$$

The Gaussian pyramid consists of a set of low pass filtered copies of the original image at different sizes. On the other hand, the Laplacian pyramids have broken down the original image into a set of bandpass images. After the image is broken down into its laplacian pyramid structure of decreasing frequencies, the main noise and useful signal components of the image exist in different layers because of their different frequency in nature. Speckle noise has high frequency and mainly exists in low pyramid layers. On the other hand, in the highest pyramid layer (i.e., Coarset scale at the top) speckle noise is negligible. Thus, performing spatial adaptive filtering in each bandpass layer can effectively remove unwanted noise from an image.

## 2.8 Modified Diffusivity Function

The second step in the proposed method is the nonlinear diffusion filtering in each of the bandpass layers of laplacian pyramid to suppress speckle noise while preserving edges. Nonlinear diffusion filtering removes speckle noise from an image by modifying the original image via a partial differential equation (PDE), Perona and Malik (1990) proposed the nonlinear diffusion as described by the following equation:

$$\frac{\partial}{\partial t} I(x, y, t) = \nabla \cdot (c(x, y, t) \nabla I) \qquad I(t = 0) = Io \tag{4}$$

where $I(x,y,t)$ is the image, $t$ is the iteration step and $c(x,y,t)$ is the diffusion function, $Io$ is the original image. For linear diffusion, the value of diffusion function $c(x,y,t)$ is constant for all image locations. For nonlinear diffusion, diffusion function in Equation (4) is a gradually decreasing function of the gradient magnitude. Thus, nonlinear diffusion diminishes across texture regions with large gradient magnitudes and is enhanced within texture regions with small gradient magnitudes. This leads to improved structure preservation and image denoising characteristics. Perona and Malik (1990) suggested that the two diffusivity functions are:

$$C_1(x,y,t) = exp\left(-\left(\frac{|\nabla I(x,y,t)|}{\lambda}\right)^2\right) \tag{5}$$

$$C_2(x,y,t) = \frac{1}{1+\left(\frac{|\nabla I(x,y,t)|}{\lambda}\right)^2} \tag{6}$$

where $\lambda$ is a constant called gradient threshold. It plays an important role in determining the degree of smoothing in the nonlinear diffusion process. The 2-D discrete nonlinear diffusion structure is translated into the following form:

$$I_{i,j}^{n+1} = I_{i,j}^n + (\nabla t).\begin{bmatrix} C_N(\nabla_N I_{i,j}^n).\nabla_N I_{i,j}^n + C_S(\nabla_S I_{i,j}^n).\nabla_S I_{i,j}^n + \\ C_E(\nabla_E I_{i,j}^n).\nabla_E I_{i,j}^n + C_W(\nabla_W I_{i,j}^n).\nabla_W I_{i,j}^n \end{bmatrix} \tag{7}$$

Subscripts *N, S, E* and *W* (North, South, East, and West) describe the direction of the local gradient, and the local gradient is calculated using nearest-neighbor differences as:

$$\nabla_N I_{i,j} = I_{i-1,j} - I_{i,j} \tag{8}$$
$$\nabla_S I_{i,j} = I_{i+1,j} - I_{i,j} \tag{9}$$
$$\nabla_E I_{i,j} = I_{i,j+1} - I_{i,j} \tag{10}$$
$$\nabla_W I_{i,j} = I_{i,j-1} - I_{i,j} \tag{11}$$

and the gradient magnitude is calculated by using Equation (12)

$$\|\nabla I\| = 0.5 \times \sqrt{\|\nabla I_N\|^2 + \|\nabla I_S\|^2 + \|\nabla I_W\|^2 + \|\nabla I_E\|^2} \tag{12}$$

To solve the Partial Differential Equation (7), a boundary condition needs to be imposed to fit the solution to an original problem. The diffusivity function defined in the literature survey has several practical limitations.

It needs a reliable estimate of image gradients because with the increase of the noise level, the effectiveness of the gradient calculation degrades and diminishes the performance of the method. Secondly, the equal number of iterations in the diffusion of all the pixels in the image leads to blurring of texture and fine edges.

The diffusivity functions in (5) and (6) may also create difficulty in dealing with convergence and high sensitivity to speckle noise present in ultrasound images. Although these problems can be solved by applying regularizing finite difference discretizations, it would be desirable to have a regularization that does not depend on discretization effects. Before reconstruction of the diffused laplacian pyramid domain, the Gaussian low pass filter is applied. In the proposed Modified LPND, a Gaussian regularization strategy which estimates the gradient ∇I on a Gaussian low pass filtered version of the image is adopted. Based on this strategy, (4) becomes

$$\frac{\partial I}{\partial t} = div[c(\|\nabla(G(\sigma)*I)\|).\nabla I] \tag{13}$$

where σ is the standard deviation of a Gaussian filter G and describes a level of uniform smoothing used to measure the image gradient, and * denotes convolution. The value of σ is chosen empirically based on the noise suppression and structure preservation.

All the above considerations suggest an approach that incorporates adaption to the ultrasound image gradient structure. The proposed modified LPND uses a gradient threshold value at each subband that is computed by using modified diffusivity function equations. It has been found that the Equation (6) has greater decay rate of the diffusivity function and will create sharper edges that persist over longer time intervals in a narrower range of edge slopes. A more gradual decay rate will sharpen edges over a wider range of edge slopes. For the same value of $\lambda$, the two diffusivity function in (5) and (6) will lead the diffusion process to significantly different results (Jin et al. 2000). To make two diffusivity functions give similar results for the same value of $\lambda$, the work modifies the modified diffusivity function given by Equation (6) as

$$C_2(\|\nabla I\|) = exp\,[1 - (\|\nabla I\|^2 / (2\lambda + 1))^2] \tag{14}$$

An improved algorithm has been adopted in the literature Fan Zhang et al. (2007). From Equation (14), at each subband in pyramid, coefficients are improved by multiplying the gradient threshold value by factor two. For nonlinear diffusion now, the diffusivity function $C_2(\|\nabla I\|)$ is a gradually increasing function of the gradient magnitude. Thus diffusion is enhanced across regions with small gradient magnitudes. This leads to improved structure preservation and preserves small detailed structures present in ultrasound liver images compared to Equation (6). Finally, reconstruction of an image from its Laplacian pyramid can be achieved by simply reversing the decomposition steps.

*Median Absolution Deviation Estimator*

The selection of the gradient threshold $\lambda$ plays a major role in determining the small parts of an image that will be blurred or enhanced in the diffusion process. Calculation of MAD estimator and its new decision role on the gradient threshold $\lambda$ for the proposed modified LPND method.

$$\lambda = \frac{MAD(\|\nabla I\|)}{0.6745} \tag{15}$$

where the constant is derived from the fact that the MAD of a zero means normal distribution with unit variance is *0.6745*.

$$\lambda(I) = \frac{1}{0.6745} \cdot MAD((\|\nabla I(l)\| \sqrt{2\log((l+1)/2l)} \tag{16}$$

where $l$ is the pyramid layer. In the proposed method, the pyramid layer is about $l = 5$. This new decision rule can help MLPND suppress noise more thoroughly and preserve important image features more effectively. If the value of $\lambda$ is set very high, the diffusion will act as a smoothing filter and cyst information is hidden by speckle. If $\lambda$ is set too low, the diffusion will act as a sharpening filter and some big noise will be preserved. However, some automatic threshold selection mechanism requires a bipeak histogram that is not common in medical ultrasound images. This book has proposed the modified gradient threshold $\lambda$ as in Equation (16) that estimates using the median absolute deviation estimator that is applied directly to remove speckles in medical ultrasound image. The entire procedure for Modified Laplacian Pyramid Nonlinear Diffusion Method for Speckle reduction on Ultrasound Liver images is shown in Figure 1.

> **Step 1:** Decomposition of a given input image into its Laplacian pyramid domain by using EXPAND and REDUCE operators.
>
> a. REDUCE: it performs 2D low pass filtering followed by a sub-sampling factor of two in both directions using the Equation (2).
> b. EXPAND: it enlarges an image into twice its size in both directions by up sampling using Equation (3).
>
> **Step 2:** Calculation of pyramid coefficient values by modified diffusivity function for nonlinear diffusion in order to remove speckles from an image.
>
> a. Apply Gaussian low pass-filter to each subband to estimate gradient $\nabla I$ by Equation (13). The selection of $\sigma$ provides noise suppression and structure preservation. Compute modified diffusivity function in Equation (14).
>
> **Step 3:** Reconstruction of the diffused laplacian pyramid by applying Gaussian low pass filter.
>
> a. MAD estimator and its new decision making role on the gradient threshold $\lambda$ for the proposed LPND is computed by Equations (15) and (16).

## Procedure for MLPND Method

## 2.9 Experimental Results

### 2.9.1 Metrics for Speckle Reduction

In this experiment, two metrics: CNR and PSNR are calculated to compare the performance of proposed MLPND method with existing methods (Nonlinear Diffusion (ND), Speckle Reducing Anisotropic Diffusion (SRAD) and Laplacian Pyramid Nonlinear Diffusion (LPND) used for speckle reduction.

The Contrast to Noise Ratio is calculated to evaluate the measure for assessing the ability of an imaging procedure to generate clinically useful image contrast. Even, if the image has a high signal-to-noise ratio, it is not useful unless there is a high enough CNR to distinguish among different tissues and tissue types and in particular between healthy and pathological tissue. The CNR, that is referred to as ultrasound lesion signal-to-noise ratio (Chen et al. 1996) is computed by Equation (13).

$$CNR = \frac{|\mu_1 - \mu_2|}{\sqrt{\sigma_1^2 + \sigma_2^2}} \quad\quad (17)$$

where $\mu_1$ and $\sigma_1^2$ are the mean and variance of intensities of pixels in a pathology bearing region (PBR), and $\mu_2$ and $\sigma_2^2$ are the mean and variance of intensities of pixels in a background region that has the same size as the PBR to be compared with. The PBR is extracted by the radiologist. The CNR for LPND is about 14.18dB and CNR value for MLPND is about 15.29 which are shown in Figure 14 and Figure 15. It can be observed from Figure 15 that the CNR for MLPND is higher and removed speckles more effectively compared to CNR for LPND in Figure 14.

Another important measure is PSNR. PSNR is the ratio between possible power of a signal and the power of corrupting noise that affects the fidelity of its representation. PSNR is a measure that is normally expressed in decibel scale which is mathematically shown in Equation (14). Higher levels of PSNR indicate that the value of useful information is greater than noise. Therefore greater values of PSNR are preferable.

$$PSNR = 20 \log_{10}\left(\frac{255}{\sqrt{MSE}}\right) \quad\quad (18)$$

$$MSE = \frac{\sum f(i,j) - F(i,j)^2}{MN} \quad\quad (19)$$

where MSE (Mean Square Error) is an estimator in many ways to quantify the amount by which a filtered/noisy image $F(i,j)$ differs from noiseless image $f(i,j)$. In this section, the MLPND using both normal and diseased ultrasound liver image data is being tested. In this experiment, the image compares the results of the MLPND with those of three existing systems, i.e., ND, SRAD and LPND filter. The CNR and PSNR are used to quantify the performance of algorithms in homogeneous regions. Initially ultrasound liver Hemangioma image is loaded in to the pre-processing system for speckle reduction which is shown in Figure 16.

The input (ultrasound liver) image is decomposed into laplacian pyramid domain of layer d = 5 and computes the gradient threshold value $\lambda$ with the support of MAD estimator. The resultant image (Image I) generated from this step is shown in Figure 17.

At each sub band in pyramid layers, laplacian filter is applied on image I and followed by REDUCE operator to generate the resultant image (Image II) that is shown in Figure 18.

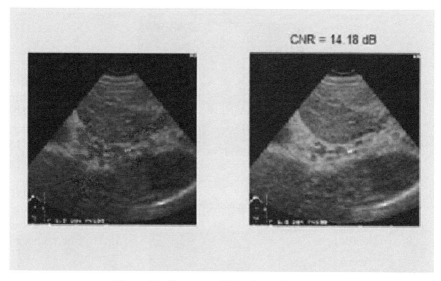

**Figure 14.** Contrast to Noise Ratio for LPND.

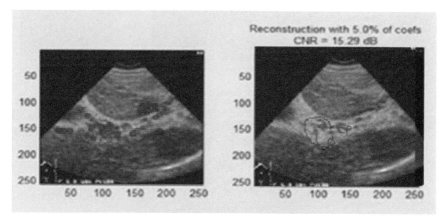

**Figure 15.** Contrast to Noise Ratio for MLPND.

Gaussian low pass filter is applied before computing diffusivity function on Image II. The resultant image is shown in Figure 19.

Anisotropic diffusion with modified diffusivity function is applied for image II that generates resultant image (Image III). It checks whether the value of k is small or not. If k value is less than its gradient threshold, the output image is clearly distinguished by its small lesion and cyst which is

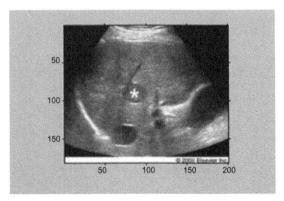

**Figure 16.** Loading Liver Hemangioma for Speckle Reduction by MLPND.

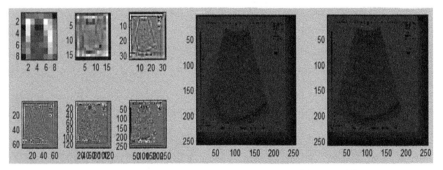

**Figure 17.** Liver Hemangioma (Image I) in Pyramid Domain (d = 5).

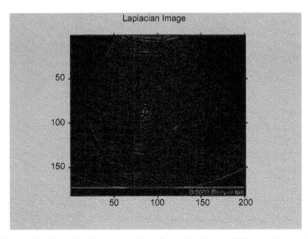

**Figure 18.** Resultant Image of Applying Laplacian Filter on Image I.

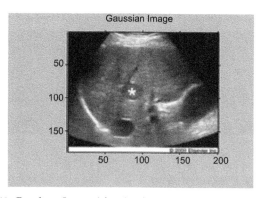

**Figure 19.** Resultant Image After Applying Gaussian Filter on Image II.

shown in Figure 20. For the same ultrasound liver image input, on applying existing filter SRAD is also shown in Figure 21.

From the above experiment, it is noticed that modified laplacian pyramid nonlinear diffusion methods works directly on ultrasound image data. The result obtained from MLPND is much sharper in terms of edge preservation and smoother in terms of speckle noise reduction than the other three filtered results as shown in Table 2 and Table 3. Table 2 and Table 3 shows evaluation results for the processed image. Also Table 2 and Table 3 highlights the comparison of CNR and PSNR average values for the liver images using MLPND with existing ND filters.

From Table 1, it is inferred that the CNR values indicate that the proposed method gives a better processing result in terms of structure preservation and contrast enhancement. For example, the CNR values for noisy cyst image is 5.34, ND and SRAD values for cyst are 11.87 and 11.36, which is comparatively similar, because ND is applied to log compressed scan data, SRAD filter is applied to uncompressed echo envelope image and CNR value for cyst image in LPND filter being 10.89, that is applied to log compressed scan data. But the proposed approach is applied directly on ultrasound image data and the value of CNR is 12.34 comparatively higher than other existing filters. From this result, it is very clear that the small subtle features like cyst are clearly differentiated by means of modified LPND method. The proposed MLPND preserves edges and small structure while maximally removing speckle and is also applied to raw ultrasound images.

From Table 2, it is inferred that the average PSNR value indicates that the proposed method is better than the other methods in terms of edge preserving ability. The PSNR values for Liver cirrhosis—noisy image is

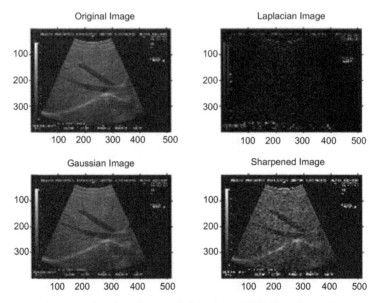

**Figure 20.** Resultant Image III of Applying MLPND on Image II.

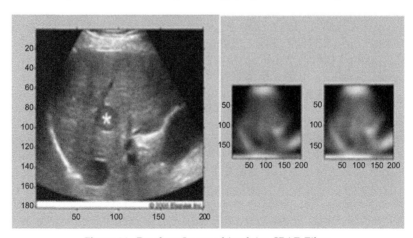

**Figure 21.** Resultant Image of Applying SRAD Filter.

**Table 2.** Comparison of CNR Values for Liver Images using MLPND with Existing Non-linear Diffusion Filters.

| Liver Diseases Filters | Normal | Cyst | Hemangioma | Hepatoma | Fatty | Cirrhosis |
|---|---|---|---|---|---|---|
| Noisy | 4.45 | 5.34 | 5.90 | 4.40 | 6.12 | 6.23 |
| ND | 9.45 | 11.87 | 17.10 | 9.18 | 9.20 | 10.92 |
| SRAD | 9.41 | 11.36 | 18.57 | 14.18 | 15.23 | 16.01 |
| LPND | 9.46 | 10.89 | 19.01 | 14.63 | 15.61 | 16.94 |
| Modified LPND | 10.84 | 12.34 | 19.22 | 15.29 | 17.52 | 18.11 |

**Table 3.** Comparison of PSNR Values for Liver Images using MLPND with Existing Non-linear Diffusion Filters.

| Liver Diseases Filters | Normal | Cyst | Hemangioma | Hepatoma | Fatty | Cirrhosis |
|---|---|---|---|---|---|---|
| Noisy | 5.49 | 4.99 | 5.89 | 5.72 | 7.13 | 8.29 |
| ND | 9.01 | 9.45 | 12.44 | 10.91 | 14.70 | 15.92 |
| SRAD | 9.25 | 8.11 | 12.56 | 13.44 | 14.71 | 14.98 |
| LPND | 10.11 | 11.89 | 13.15 | 14.99 | 15.74 | 17.36 |
| Modified LPND | 11.45 | 13.32 | 15.55 | 16.58 | 17.26 | 18.35 |

about 8.29. The LPND filter produces better results compared to SRAD and ND, whereas the proposed method produces higher PSNR values for liver cirrhosis. Due to its optimal noise reduction, MLPND provide better results compared to the other three methods, resulting in increased contrast and improved visibility of small structures (i.e., subtle features like lesion) in each image. The result has been investigated by the physician Dr. P.S. Rajan, GEM, Coimbatore and Dr. Mahalakshmi, S, MMH, Madurai and it has been concluded that speckle found in the PBR is removed from ultrasound images by leaving its semantic content.

These quantitative results show that the MLPND can eliminate speckle noise without distorting useful image information and without destroying the important image edges. All the other three methods have limited noise reduction performance. SRAD broadens the boundaries of bright regions and shrinks those of dark regions in ultrasound images. Although Nonlinear Diffusion (ND) enhances the coherence of organ surfaces, LPND also causes blurring of small regions, whereas MLPND produce clear edge details and preserves small subtle features like lesion and cyst. From the above two Tables 1 and 2, it is inferred that the speckle is removed and the edge details are also restored effectively. The CNR value is able to provide conclusion that the image clarity is not affected.

# 3

# Image Registration

## 3.1 Introduction

Medical images are being increasingly widely used in healthcare and in bio-medical research and a very wide range of imaging modalities, such as CT, MRI, PET, SPECT, and so on is now available. In some clinical scenarios, the information from several different imaging modalities should be integrated to deduce useful clinical conclusions. Image registration aligns the images and so establishes correspondence between different features contained on different imaging modalities, allows monitoring of subtle changes in size or intensity over time or across a population and establishes correspondence between images and physical space in image guided interventions. Image registration is the process of determining the spatial transform that maps points from one image (defined as the moving image) to homologous points on an object in the reference image (called as the fixed image). The similarity of the two images will be calculated and investigated after each transform until they are matched. Of the multitude of image registration similarity measures that have been proposed over the years, mutual information is currently one of the most intensively researched measures. This attention is a logical consequence of both the favorable characteristics of the measure and the good registration results reported. Mutual information is an automatic, intensity-based metric, which does not require the definition of landmarks or features such as surfaces and can be applied in retrospect (Amores and Radeva 2005). Furthermore, it is one of the few intensity-based measures well suited to registration of multimodal images. Unlike measures based on correlation of gray values or differences of gray values, mutual information does not assume a linear relationship among the gray values in the images. However, the mutual information registration function can be ill-defined, containing local maxima. Another drawback of MI is that it is calculated on a pixel-by-pixel basis, applying that it takes into account only the relationship between corresponding individual pixels and not those of each pixel s respective neighbourhood. As a result, much of the global spatial information inherent in images is not utilized. In addition, it is a time-consuming work, especially for high resolution images, because mutual information of the two images must be calculated in each iteration. In this chapter, we develop a new method for automatic registration of head images by computer, which obtained CT and MR images employing maximization of mutual information, and reduce the processing time.

Registration of medical images is done to investigate the disease process and understand normal development and aging. To analyze the image from different scanners, all the images need to be aligned into the same location, where the structure of tissues can be compared. Patient information is accumulated over years and has details in the form of ultrasound images. This needs to be stored and retrieved when required. Traditional mechanism for storage and retrieval suffers from redundancy that could be eliminated during image registration. In modern radiology, image registration is important for diagnosis, surgical planning and treatment control.

For classification and retrieval of ultrasound liver images, image registration helps to reduce the computational time and improve the accuracy of retrieval for diagnosis purposes. Mutual information is a popular similarity measure for medical image registration. This chapter focuses on the registration of ultrasound mono-modal liver images and comparison is done by means of optimization techniques using mutual information and is very well suited for clinical applications. Image registration reduces the redundancy of retrieval images and increases the response rate that motivates this work.

## 3.2  Importance of Medical Image Registration

The primary goal of medical image registration is the process of geometrically aligning different sets of medical image data into one coordinate system. This alignment process requires the optimization of similarity measures. Mutual information (MI) is a popular entropy-based similarity measure which has found use in a large number of image registration applications. Stemming from information theory, this measure generally outperforms most other intensity-based measures in multimodal applications, as it does not assume the existence of any specific relationship between image intensities. It only assumes a statistical dependence. In this work, MI is selected in mono-modal applications to monitor the evolution of liver pathology diseases. The basic concept behind any approach using mutual information is to find a rigid body transformation, when applied to an image will maximize the MI between two images. While rigid body registration has become a widely used tool in clinical practice, non-rigid body registration has not yet achieved the same level for the surgeon during any decision making processes. Major difficulties in image registration process are image noise, illumination changes and occlusions. The examples for medical image registration is shown in Figure 22.

As mentioned above, in this book, focus is on using mono-modal image registration for monitoring the growth of liver diseases from benign to malignant by matching ultrasound images of the same patient taken at different times and also for classification of ultrasound liver images. There are many medical image registration methods, and they may be classified in varied ways. Brown (1992) proposed a classification for general registration techniques which included the feature space, similarity metric, search space, and the search strategy. Maintz et al. (1998) has suggested a nine-dimensional scheme that provides an excellent categorization. However, the set of criteria that is supported for this Book includes image alignment

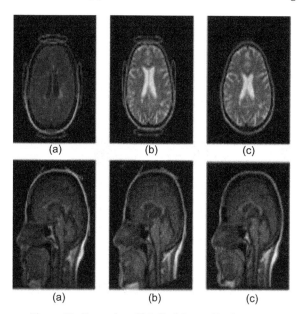

**Figure 22.** Examples of Medical Image Registration.

algorithm, geometric transformation, domain methods, optimization procedure, modalities involved, and automation models as shown in Table 4.

Image registration involves the mapping of coordinates between any two images by means of same features (point, lines, surfaces or intensity values) in the two images that are mapped to each other. The main components of image registration technique are:

i) Similarity measures to determine the quality of match of the two images.

**Table 4.** Classification of Medical Image Registration Techniques.

| Based on Image Alignment Algorithm | |
|---|---|
| Intensity based image registration | It compares intensity patterns in images via correlation or mutual information metrics. It registers entire image or sub image. |
| Feature based image registration | This method finds correspondence between image features such as points, lines, and contours. It establishes a correspondence between a numbers of especially distinct points in images. |
| **Based on Transformation Models** | |
| Rigid-body transformation | Also called linear transformations, which includes rotation, scaling, and other affine transforms. |
| Non-rigid transformation | Also called elastic transformation, which includes radial basis functions, physical continuum models, and large deformation models. |
| **Based on Domain Methods** | |
| Spatial domain methods | Spatial methods operate in the image domain, by matching intensity patterns or features in images by using control points. |
| Frequency domain methods | Frequency-domain methods find the transformation parameters for registration of the images while working in the transform domain, such as translation, rotation, and scaling. |
| **Based on Modalities** | |
| Mono-modal | Single-modality methods tend to register images in the same modality acquired by the same scanner/sensor type. |
| Multi-modal | Multi-modality registration methods tend to register images acquired by different scanner/sensor types. |
| **Based on Automation Methods** | |
| Automatic image registration methods | Automatic methods provide tools to align the images itself without any user. |
| Interactive image registration methods | Interactive methods reduce user bias by performing certain key operations automatically while still relying on the user to guide the registration. |

ii) Transformations that relate the two images. The most representative transformation models include rigid, affine, projective and curved.
iii) Optimization techniques to determine the optimal value of the transformation parameters as a function of the similarity measure.

From the application s point of view, registration algorithms can be classified based on several criteria. The classifications presented here are partially based on (Mousavi et al. 2014). The criteria and their primary subdivisions are:

- Modality: Mono-modal refers to the case where all images are obtained from the same imaging sensor type and there are no major differences between the intensity ranges that correspond to the same physical/physiological phenomenon. In a multi-modal setting, these ranges can differ drastically. This is typically due to different sensor types.
- Dimensionality: This refers to the number of dimensions of the images. Historically, images have typically had two spatial dimensions. Today, however, several imaging technologies provide 3D volumes. Moreover, some sensors, e.g., functional MRI, provide a video, i.e., a sequence of images. When treating the video as one big data set, time can be thought of as an extra dimension. One convention is to denote time as a 0.5 dimension. This is helpful to clarify some ambiguities, e.g., 3D (three spatial dimensions) versus 2.5D (two spatial dimensions + time). Most of today's applications involve 2D/2D and 3D/3D registration.
- Speed: Offline refers to applications where time is not an important constraint. Online denotes a heavy time constraint, typically indicating real-time applications. An important online example is intra-operative procedures performed within the operating theater. Some scientific applications (e.g., human brain8 mapping), on the other hand, do not have a heavy time constraint.
- Subject: In a medical application, intra-subject refers to the task where all images are of the same subject (patient). Inter-subject denotes the fact that more than one subject is involved. If an averaged template (atlas) is employed, this is typically called atlas registration. Inter-subject applications are typically more complex, since correspondence is difficult to identify.
- Nature of Misalignment: Geometric misalignment can be attributed to several factors, including different viewpoints (orientation) of sensors, temporal changes (e.g., Digital Subtraction Angiography:

the registration of images before and after radio isotope injections to characterize functionality) and inherent differences (e.g., brains of different subjects).

From the methodology point of view, registration algorithms can be classified based on several criteria:

- Employed Information Content: In the registration literature, one can identify two trends in the type of information employed. Landmark based approaches rely on the definition of landmarks. Alignment is computed based on these landmarks (sets of points, lines or surfaces) only. These landmarks can have a clear physical meaning (e.g., the cortical surface of the human brain (Medina et al. 2012), fiducial markers visible in all modalities (Singh et al. 2014), etc.), or they can be of theoretical interest only (e.g., lines, corners, points of high curvature, etc.). In landmark based registration, the set of identified points is sparse compared to the original image content, which allows fast optimization. However, performance of the algorithm heavily depends on the landmark identification. Image content based approaches, on the other hand, rely on pixel intensity information. These typically extract features from pixels (e.g., intensity values (Cox et al. 1997), gradient vectors (Zicari 2012), wavelet coefficients (Zhao and Bose 2002), etc.) and compute an alignment based on the set of feature samples. These are usually slower than landmark based algorithms, but have the potential to produce accurate and robust results in contexts where landmarks are difficult to define or determine.
- Locality of Alignment Measure: Alignment quality can be measured for the whole image, using global measures, e.g., sum of squared differences of all pixel values, or for a neighborhood of a pixel location using local measures, e.g., local correlation.
- Transformation: Generally speaking, there are two types of geometric transformations: parametric models, e.g., rigid-body, affine, spline based, etc., where a small set of parameters determine the transformation and nonparametric models (also known as optical flow, dense matching, etc.), where each pixel is allowed to move independently. Note that in the latter case, if there was no restriction on the transformation, an image could be made to look similar to any other image with the same intensity range as the first image. Thus, these methods require regularization to overcome mis-alignment and incorporate prior knowledge about the deformation field.

- Optimization: Typically, iterative methods are employed within a multiresolution pyramid, to speed up convergence. Popular choices of optimizers are: gradient-descent and its variants (Cox et al. 1997), Powell's method (Studholme and Hawkes 2000), Downhill simplex method and Levenberg-Marquardt optimization (Tian and Huang 2000). In this thesis, we restrict our analysis to global image content-based approaches, which provide a general framework and require minimal knowledge about the specifics of the application domain.

## 3.3 Intensity Vs Feature Based Image Registration

Medical image registration can be characterized based on the image information into intensity-based and feature-based methods. The first method utilizes an image intensity to estimate the parameters of a transformation between two images using an approach involving all pixels of the image. In contrast, the second method does not work directly with image intensity values and relies on establishing feature correspondence between the two images. The feature-based matching algorithm may be performed by iterative closest point algorithms. The above methods use feature matching techniques to determine corresponding feature pairs from the two images and then compute the geometric transformation relating them. For rigid body transformation, typical approaches include landmark-based registration, surface based registration, and intensity based Registration. The drawbacks with landmark based registration and surface based registration is resolved by suggesting the intensity based image registration technique which in this work is supported to register ultrasound liver images.

Intensity based registration methods, also referred to as voxel property based methods, is significantly different from segmentation based registration methods as shown in Figure 23. This method operates directly on the intensity values within the image and thus do not need to utilize complex segmentation procedures/other feature extraction methods in order to obtain features required for matching. These approaches are generally the most flexible and robust of all registration methods.

## 3.4 Similarity Measures

The best choice of a similarity measure is heavily dependent on the underlying images to be registered. A similarity measure should have its global maximum when the two images are correctly aligned. There are several similarity measures that are available for medical image registration.

They are Maximum-Likelihood Formulation, Sum of Squared Differences (SSD), Normalized Cross-Correlation (NCC) also denoted as Correlation Coefficients (CC), Correlation Ratio (CR) and Mutual Information (MI). CC is commonly used for mono-modal image registration to access the amount of linear correlation of two signals by computing the average product of its demeaned values, divided by their standard deviation. These methods were introduced to help overcome the problem of differing levels of intensity values between images. Correlation ratio is a similarity measure but not symmetric. This work compares mono-modal liver image registration by CC and MI.

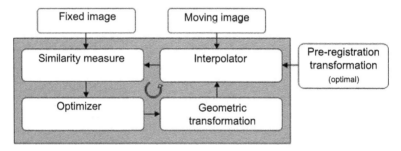

**Figure 23.** Intensity Based Image Registration.

### 3.4.1 Correlation Coefficient (CC)

The Correlation coefficient (CC) is optimal, when the dependency between the two ultrasound liver images intensities is linear. This is a reasonable hypothesis in case of mono-modal image registration. This coefficient measures the dispersion of the joint intensities along a line. The correlation coefficient is mathematically described as follows:

$$\rho = \frac{E((I - \mu_I)(T(J) - \mu_{T(J)}))}{\sigma_I \sigma_{(J)}} \tag{20}$$

where I and J are the intensities of the images, T(J) is the image J transformed by T, $\mu_I$ and $\mu_T(J)$ are the mean intensities of I and T(J) and $\sigma I$ and $\sigma T(J)$ are the standard deviations of I and T(J).

### 3.4.2 Mutual Information (MI)

The most successful automatic image registration methods are based on MI, which is a measure of the dependence between two images. Mutual

Information (MI) is a popular entropy-based similarity measure, that has found use in a large number of image registration applications and it is suited for all modalities of image registration. Moreover, it is one of the few intensity-based measures. Mutual information (MI) is a statistical measure that finds its roots in information theory (Pluim et al. 2000). MI is a measure of how much information one random variable contains about another. The idea behind Mutual information (MI) is to align two images correctly by minimizing the amount of information in a shared representation. This similarity measure is supported in this work. Mutual information has the following properties (Aczel and Darocz 1975):

1. I (A, B) = I (B, A)   It is symmetric; otherwise it would not be mutual information. However, it is a logical property in theory; mutual information is not symmetric in practice. Implementation aspects of a registration method, such as interpolation and number of samples, can result in differences in outcomes when registering A to B or B to A.
2. I (A, A) = H (A)   The information image A contains about itself is equal to the information (entropy) of image A.
3. I(A, B) ≤ H(A),
   a) I (A, B) ≥ H (B)   The information that the images contain about each other can never be greater than the information in the images themselves.
   b) I(A,B) ≥ 0   The uncertainty about A cannot be increased by learning about B.
4. I (A, B) = 0   if and only if A and B are independent. When A and B are not related in any way, no knowledge is gained about one image when the other is given.

## 3.5 Geometric Transformation

The choice of the geometric transformation model is crucial to the success of a registration algorithm and is highly dependent on the nature of the data to be registered. Usually the geometric transformation is divided into rigid and non-rigid classes. The rigid transformation is the simplest one, and it can be defined by 6 parameters or degrees of freedom: 3 translation and 3 rotational parameters. In rigid transformations, the distance between corresponding points are preserved. The non-rigid transformation class includes the similarity transformation (translation, rotation and scaling), affine transformation map straight lines to straight lines while the parallelism is preserved. Projective transformations map straight lines to straight lines but parallelism is in general not preserved and the curved transformation is also commonly referred to as deformable, elastic or fluid transformation (Francisco et al. 2012). A rigid transformation involves only rotation and translation. It is the most important geometric transformation among all registration techniques. This has supported the work using MI based registration followed by rigid body transformation.

## 3.6 Mono-modal Verses Multimodal

Another classification of image registration can be made between mono-modal and multimodal methods. Monomodal methods tend to register images in the same modality acquired by the same scanner/sensor type, while multi-modality registration methods tended to register images acquired by different scanner/sensor types. Multi-modality registration methods are often used for image fusion in medical imaging as images of a subject are frequently obtained from different scanners. Examples include registration of brain CT/MRI images or whole body PET/CT images for tumor localization.

## 3.7 Optimization Techniques

The core of a medical image registration algorithm is an optimization framework involving a search of those parameters of the transformation model that minimizes the cost function. The goal of the optimization algorithm used is to search for the maximum or minimum value of the similarity measure adopted. In non-rigid registration designing an optimizer can be difficult than the rigid registration, because the more non-rigid (or flexible) the transformation model, the more parameters are generally required to describe it. So the optimizer takes a large amount of time to determine the parameters and due to local minima problem there is more chance of choosing a set of parameters that results in a good image match which is nevertheless not the best one. For optimization of image registration, Nelder-Mead Simplex (NMS), Powell s direction set method, Genetic algorithm (GA), Dividing RECTangles (DIRECT) methods, and Particle swarm optimization (PSO) method are used. Collignon et al. (1995) proposed a Powell optimization algorithm to search for the best alignment of data. This optimizer is based on a series of line minimizations in the parameter space and suffers from sensitivity to the initial order in which the parameters are optimized.

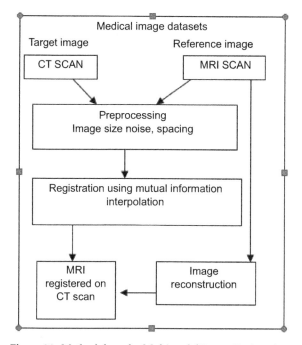

**Figure 24.** Methodology for Multimodal Image Registration.

## 3.8 Mutual Information Based Image Registration

From the above survey, it reveals that the computation time is a major issue during image registration. MI based image registration is proposed on mono modal images and further optimization helps to reduce computation time. The basic idea behind this research approach is to use mutual information based liver image registration to reduce the redundancy of retrieval images and monitoring the growth of the liver disease that motivates this work. The MI technique however is more flexible and much more robust than other similarity measure, such as correlation coefficients and correlation ratio. Growth monitoring of diffuse liver into cancer is a key domain for mono-modal registration in research perspective. The overall systematic diagram for proposed Mono-modal image registration by Mutual Information is shown in Figure 25. It includes four phases: Image Pre-processing, Similarity measure, Geometric transformation, and Optimization algorithms.

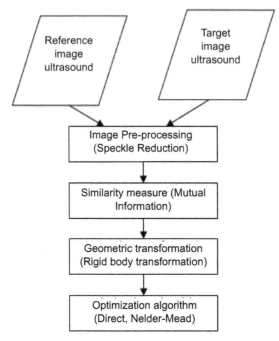

**Figure 25.** Systematic Flow diagram for MM-MI based Image Registration.

The following steps involved in MM-MI based image registration are:

1. Image Pre-processing for speckle reduction
   Pre-process both reference image (ultrasound- Liver) and target image (ultrasound-Liver) of the same patient by using S-Mean filter, i.e., (SRAD and Median filter) in order to remove speckle noise present in the images.

2. Similarity measures
   Calculate MI value by entropy measure, that is necessary for intensity based image registration for both target and reference image.

3. Geometric transformation
   Perform rigid body transformation for target image that involves translation, rotation and affine which helps to align source image.

4. Optimization techniques
   Apply Optimization techniques on rigid body—DIRECT and Nelder-Mead method to optimize the registration at minimum value. The optimization algorithm is used to search for the best alignment data.

### 3.8.1 Mono-Modal Liver Image Registration by MI

Before applying the mono-modal image registration, calculate entropy value for both pre-processed target image (ultrasound) and the reference image (ultrasound). For any probability distribution, entropy exist that has many properties that agree with the intuitive notion of what a measure of information should be. This notion is extended to define mutual information that is a measure of the amount of information one random variable contains about another. Entropy then becomes the self-information of a random variable. Mutual information is a special case of a more general quantity called relative entropy that is a measure of the distance between two probability distributions. The Shannon entropy (Shannon 2001) for joint distribution can be defined as:

$$-\sum_{i,j} P(i,j) \log P(i,j)$$

where P (i,j) is the joint probability mass function of the random variables i and j respectively.

This section focuses on the common state-of-the-art intensity-based similarity measure, Mutual information (MI). To simplify the computation of each similarity measure, the measures are computed by means of a joint histogram of the images. The construction of the joint histogram is well described by Maes et al. and shown below in Figure 26.

### a) Mutual Information

Mutual information is an automatic, intensity-based metric (Josien et al. 2003). It is one of the few intensity-based measures well suited for registration of both mono-modal and multimodal images. The mutual information similarity measure is based on the information theory and is related to the entropy of the joint histogram. Unlike measures based on correlation of gray values or differences of gray values, mutual information does not assume a linear relationship among the gray values in the images. MI is a measure of how much information one random variable contains about another. Mutual information (MI) is a statistical measure that finds its roots in information theory (Aczel and Daroczy 1975). Maes et al. (1997) computed MI based on the marginal and joint image intensity distributions P(I), P(T(J)) and P(I, T(J)), I and J being the images, T(J) the image J transformed by T and can be written mathematically as follows:

$$MI_T(I,J) = \sum_{I,J} P\left(I,T(J)\log_2 \frac{P(I,T(J))}{P(I),P(T(J))}\right) \tag{21}$$

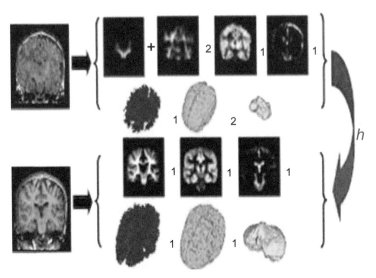

**Figure 26.** MI based Image Registration.

The MI of two random variables i and j can be defined as

$$\sum_{i,j} P(i, j) \log P(i, j) \tag{22}$$

where P(i,j) is the Joint Probability Mass Function (JPMF) of the random variables A and B, and $P_A(a)$ and $P_B(b)$ are the marginal probability mass function of A and B, respectively. The MI can also be written in terms of the marginal and joint entropy of the random variables A and B as follows:

$$I(A,B) = H(A) + H(A) \quad H(A,B) \tag{23}$$
$$= H(A) \quad H(A/B)$$
$$= H(A) \quad H(B/A) \tag{24}$$

where H(A) and H(B) are the entropies of A and B, respectively then the joint entropy is H(A,B). H (A/B) and H (B/A) is the conditional entropy of A given B and B given A respectively which is computed from the Equation (23). A distribution with only a few large probabilities has a low entropy value; the maximum entropy value over a finite interval is achieved by a uniform distribution over that interval. The entropy of an image indicates how difficult it is to predict the gray value of an arbitrary point in the image. MI is bounded by cases of either complete dependence or complete independence of A and B, yielding values of I = H and I = 0 respectively, where H is the entropy of A or B.

The strength of the MI similarity measure in mono modal image registration lies in the fact that no assumptions are made regarding the nature of the relationship between the image intensities in both target and reference image, expect that such a relationship exists. This is not the case for correlation methods that depend on a linear relationship between image intensities. For image registration, the assumption is that maximization of the MI is equivalent to correctly registering the images. It is clear from Equation (23) that if the joint entropy of A and B are not affected by the transformation parameters, maximizing the MI is equivalent to minimizing the joint entropy. It is a measure of dependence between the gray values of the images when they are correctly aligned. Mis-registration will result in a decrease in the measure.

### b) Geometric Transformation: Rigid Body Transformation

The Rigid body transformation includes rotations, translations, reflections, or their combination. Sometimes reflections are excluded from the definition of a rigid transformation by imposing that the transformation also preserves the handedness of figures in the Euclidean space preserve handedness. Fookes and Bennanoun (2002) discussed the usage of MI for rigid body medical image registration. In general, any proper rigid transformation can be decomposed as a rotation followed by a translation, while any rigid transformation can be decomposed as an improper rotation followed by a translation. Any object will keep the same shape and size after a proper rigid transformation, but not after an improper one.

The process of image registration refers to the procedure of geometrically aligning the coordinate system of two or more images. This work concentrates only on the registration of two images. Given two images, one image is selected as the reference image $R$, and the other is selected to be the floating $F$. The floating image $F$ is transformed by some linear transformation until it is spatially aligned with the reference image $R$. Let $T$ be a linear transformation with the vector parameter $\alpha$. The number of elements in $\alpha$ determines the degrees of freedom. When MI is applied to rigid and affine image registration, information from the whole image is taken into account to steer a limited set of parameters. For this 2-D application, an affine transformation with six degrees of freedom is adequate to perform the registration. Let $I(X)$ be an input liver image, after applying $Rx'$ rotation to reference image by degree two, and apply small transformation $t'$ to register reference image onto target image. Translations and rotations suffice to register images of rigid objects. Rigid registration is also used to approximately align images that show small changes in object shape or small changes in object intensity as shown in

$$f(X) = R(I(X)) + t \tag{25}$$

### 3.8.2 Multimodal Liver Image Registration

*a) Mutual Information and Image Registration*

In this extended work, we have developed a multi-modal image registration technique, which has been based on CT and MR imaging of the head. When an image superposes onto another one that has obtained multi-modality then several preprocessing techniques are required for image registration. For multi-modal image registration, the relation between the sizes of image is usually not same. In the preprocessing step, we remove some noise and normalize the size of image. In the next step, the initial point is detected by using the center of gravity of each image as an initial registration. Finally, the registration is performed employing maximization of mutual information. The details of each method are shown as follows.

*b) Preprocessing and Initial Registration*

The images obtained from different modalities may have different image size, pixel spacing and a number of slices. In the preprocessing step, image size, pixel spacing is normalized and image noise is reduced for registration. To align the multi-modal images, center of gravity is used for initial registration. In this chapter, we assume that the MR image is the target image and the CT image is the reference image. In this step, the center of gravity on multi-modal images has been calculated as follows. Let a digital image be denoted by f(x, y) where x and y are bounded positive integers. The $(i+j)^{th}$ order moment mij of f(x, y) is defined by

$$Mij = \sum x^i y^j f(x,y) \qquad (x,y) \varepsilon R$$

Where R is a specified region. The centroid denoted by $(x_c, y_c)$ is calculated by

$$(x_c, y_c) = (m_{10}/m_{00}, m_{01}/m_{00})$$

The target image is transformed onto the x y coordinates of the derived center of gravity. It must be the same coordinates as an initial position for registration. The most popular technique of interpolation is linear interpolation, which defines the intensity of a point as the weighted combination of the intensities of its neighbours.

## c) Summation of Intensity Projection with Weight

In order to reduce the computational time, three dimensional (3-D) image data is projected as a two dimensional (2-D) image data. It is similar to the Maximum Intensity Projection (MIP) method which is applied in the medical imaging field as a 3-D image representative technique. We make the six 2-D image from the 3-D image set for image registration by calculating the Summation of Intensity Projection (SIP) with weight on each direction. The weight depends on the distance from the screen. $X_k$ (k = 1,2,   n) is assumed as the change in signal density in 3-D space. The expression of SIP with weight is described as follows.

$$S = \sum_{K=1}^{N} x_k w_k$$

## d) Final Registration

Final registration is performed by using the mutual information method. For two images A and B, mutual information I(A,B) can be defined as follows (Zhang et al. 2007).

$$I(A,B) = \Sigma p_{A,B}(a,b) \log p_{A,B}(a,b)/p_A(a) \cdot p_B(b)$$

The interpretation of this equation measures the distance between the joint distribution of image pixel value $p_{A,B}(a,b)$ and the joint distribution in case of independence of the image, $p_A(a)$, $p_B(b)$. In this chapter, the target image is transformed for maximization of mutual information. When mutual information is calculated in the images, the intensities of corresponding points of two images a and b are used for making 2-D histogram h(a,b). It can be thought that the probability distribution function can be obtained by dividing the histogram by N, where N is a number of pixels.

When mutual information is detected, the intensity of pixel is linearly converted from 0 to 255. In this chapter, the amount of the movement of the target image is described by optimization that uses the multi-dimensional direction set method (powell s method) of which the index is mutual information. This method can be used for 3-D data. But, it needs more computational time. Thus the calculation cost is decreased by using 2-D data of the target image and the reference image, 2-D data is obtained by the method of SIP with weighted value.

The target image is the CT image and MRI is the reference image. When the reference image compares to the target image we get the matched image and rotated image shown in Figure 2, and Figure 3.

## Proposed Algorithm

PROCEDURE: Image Registration that uses 2-D data
Input: F and R
Output: Registration using Mutual information
Begin
Flag $\leftarrow$ 0 and $n_i \leftarrow$ 0 (i = 1,2, ..6)
$I_0 \leftarrow$ I(F,R).
/*The 2D data $f_j$ and $r_j$ are made from F and R.*/
if $n_i \leftarrow 0, j \leftarrow i$
/*The movement parameter $m_i$*/
$m_j \leftarrow f_j * r_j$.
$F_j \leftarrow$ T($F_i, m_j$), $I_j \leftarrow$ I($F_j,R$)
/*The maximum value of $I_{max}$*/
$I_{max} \leftarrow I_j$ and $I_0$.
If $I_{max} = I_j$.
$K \leftarrow j *_{Imax}$.
F = $F_k$, $n_k$ = 1, $I_0 = I_{max}$, Flag = 0.
/*if $I_{max}$ not included in $I_j$*/
If Flag = 1,
Flag = 0, $I_0 = I_{max}$ and $n_i$ = 0.

## Experimental Results

Our new technique was applied to the two different modalities images-CT and MR images of five human head images and satisfactory results were achieved. The parameters of the CT and MR images are shown in Table 5. In this paper, the image that has 512*512 pixels in each slice were converted into 256*256 pixels to reduce the processing time. The experimental results are shown in Figure 2, Figure 3, and Table 2 shows the transformation and matching of the image using mutual information.

Multi-modal medical intensity registration is an important capability for finding the correct geometrical transformation that brings one intensity pixel in precise spatial correspondence with another intensity pixel. Multi-modal intensity registration is needed in order to benefit from the complementing information in the medical images of different modalities. The result demonstrates that our registration technique allows fast, accurate, robust and completely automatic registration of multimodality medical images. From the image the value of mutual information in MRI and CT

**Table 5.** Image Parameter.

|  | CT Image | MRI Image |
|---|---|---|
| Size [pixels] | 512*512 | 512*512 |
| Pixel spacing [mm] | 0.638 | 0.638 |
| Slice thickness [mm] | 2 | 2 |

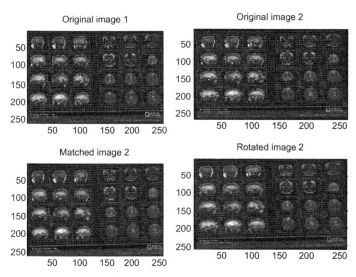

Multimodal Image Registration.

**Table 6.** Experimental Results for Multi-modal Image Registration.

| Number | Existing Method | Proposed Method |
|---|---|---|
|  | Mutual Information | Mutual Information |
| 1 | 0.4473 | 0.5658 |
| 2 | 0.5174 | 2.5083 |
| 3 | 0.4667 | 2.5396 |
| 4 | 0.5639 | 2.9598 |

are compared with the existing method and shown in Table 6. Compared with the existing method, our experimental results are better by using the intensity of two images.

## 3.9 Optimization Techniques

The primary goal of medical image registration is the process of geometrically aligning different sets of medical image data into one coordinate system. This alignment process requires the optimization of similarity measures. This work involves two optimization techniques namely DIRECT and Nelder-Mead method.

### 3.9.1 Optimizing MI by Direct Method

The DIRECT technique comes from the shortening of the phrase "Dividing RECTangles , which describes the way the algorithm moves towards the optimum. DIRECT is a sampling algorithm. That is, it requires no knowledge of the objective function gradient. Instead, the algorithm sample points in the domain, uses the information it has obtained to decide where to search next. A global search algorithm like DIRECT can be very useful when the objective function is a black box function or simulation.

The algorithm begins by scaling the design box to an n-dimensional unit hypercube. The centre point of the hypercube is evaluated and then points are sampled at one-third the cube side length in each coordinate direction from the centre point. Depending on the direction with the smallest function value, the hypercube is then subdivided into smaller rectangles with each sampled point becoming the centre of its own n-dimensional rectangle or box. All boxes are identified by their centre point $c_i$ and their function value $f(c_i)$ at that point.

From there, the algorithm loops in a procedure that subdivides each of the boxes in the set in turn until termination or convergence. By using different value of the Lipschitz constant, asset of potentially optimal boxes is identified from the set of all boxes. These potentially optimal boxes are sampled in the directions of the maximum side length, to prevent boxes from becoming overly skewed, and subdivided again based on the directions with the smallest function value. If the optimization continues indefinitely, all boxes will eventually be subdivided meaning that all regions of the design space will be investigated. The algorithm for DIRECT Optimization method is shown below.

---

**Algorithm—DIRECT Optimization Method**

1. Normalize the search space to be the unit hypercube. Let $C_1$ be the center point of this Hypercube and evaluate $f(C_1)$.
2. Identify the set S of the potentially optimal rectangles (those rectangles defining the bottom of the convex hull of a scatter plot of rectangle diameter versus $f(C_i)$ for all rectangle centers $C_i$.
3. For all rectangles j€s:

   3a. Identify the set I of dimensions with the maximum side length. Let $\delta$ equal one third of this maximum side length.

   3b. Sample the function at the points $c - \delta e_i$ for all i€I, where c is the center of the rectangle and $e_i$ is the $i^{th}$ unit vector.

   3c. Divide the rectangle containing c into thirds along the dimensions in I, starting with the dimension with the lowest value of $f(c - \delta e_i)$ and continuing to the dimension with the highest $f(c - \delta e_i)$.
4. Repeat 2 3 until stopping criterion is met.

---

## Algorithm for DIRECT Optimization Method

### *3.9.2 Optimizing MI by Nelder-Mead Method*

The Nelder-Mead Simplex (NMS) or Downhill simplex search method is based upon the work of Spendley et al. (1962). Nelder-Mead generates a new test position by extrapolating the behavior of the objective function measured at each test point arranged as a simplex. Nelder-Mead simplex method does not require any calculation of derivatives. A NMS method tries to crawl into global minimum of several variables which has been devised by Lagarias (Lagarias et al. 1998). One foot is the starting estimate, and the others are generated within the search space, either randomly or in a structured fashion. Because the NMS is the simplest creature that requires N dimensions to exist, all the remaining feet lie in an $N$  1 dimensional plane. The creature first tries to reflect its worst foot across this plane (reflection). If the new position is better than the one it came from, it tries to go further in that direction (expansion). If the position after reflection is worse, it tries to place its foot in an intermediate position (contraction). If none of these moves gives a better solution than the worst leg, it tries moving all of its feet closer to the best one. This procedure continues until the stopping criterion is reached. It is effective and computationally compact. The algorithm for Nelder-Mead Optimization method is shown below.

---

### Algorithm—Nealder-Mead Optimization Technique

1. Evaluate f at the vertices of S and sort the vertices of S so that $f(x_1) \le f(x_2) \le \ .. \le f(x_{N+1})$ holds.

2. Set $f_{count} = N + 1$.

3. while $f(x_{N+1}) \quad f(x_1) > \tau$

    a) Compute x, $x = \frac{1}{N}\sum_{i=1}^{N} x_i$, $x(\mu_r)$, $x_\mu - (1 + \mu)x \quad \mu x_{N+1}$ and $f_r = f(x(\mu_r))$. $f_{count} = f_{count+1}$

    b) Reflect: if $f_{count} = K_{max}$ then exit. If $f(x_1) \le f(x_r) \le f(X_N)$ replace $X_{N+1}$ with $x(\mu_r)$ and step 3g.

    c) Expand: if $f_{count} = K_{max}$ then exit. If $f(x_r) \le f(x_1)$ then compute, $f_e = f(x(\mu_e)).f_{count} = f_{count}+1$. If $fe < f(x_r)$, replace $x_{N+1}$ with $x(\mu_e)$, otherwise replace $x_{N+1}$ with $x(\mu_e)$. goto step 3g.

    d) Outside contraction: if $f_{count} = K_{max}$ then exit. If $f(x_N) \le f(x_r)) \le f(X_{N+1})$, compute $f_c = f(x(\mu_{oc})).f_{count} = f_{count}+1$. If $f_c \le f(x_r)$ replace $x_{N+1}$ with $f(x(\mu_{oc}))$ and goto step 3 otherwise goto step 3f.

    e) Inside contraction: if $f_{count} = x_{max}$ then exit. If $f(x_r) \ge f(x_{N+1})$, compute $= f(x(\mu_{ic}))$. If $f_c \le f(x_{N+1})$ replace $(x_{N+1})$ with $f(x(\mu_{ic}))$ and go to step 3g. Otherwise goto step 3f.

    f) Shrink if $f_{count} \ge x_{max} \quad N$ then exits. For $2 \le I \le N + 1$: set $x_i = x_1 \frac{(x_i - x_1)}{2}$ -; compute $f(x_i)$.

    g) Sort: sort the vertices of S so that $f(x_1) \le f(x_2) \le \ .\le f(x_{N+1})$ holds.

---

## Algorithm for Nelder-Mead Optimization Method

The initial simplex is important, indeed, a too small initial simplex can lead to a local search and consequently the NM can get more easily stuck. So this simplex should depend on the nature of the problem. Both DIRECT and Nelder-Mead optimization techniques are tested to determine the minimum computation time during MM-MI image registration. The performance of the two optimization techniques namely DIRECTS and Nelder-Mead method have been compared. It shows that the Dividing Rectangles (DIRECT) method can quickly produce optimal solutions that reduce the computation time of the registration compared to Nelder-Mead method. This overcomes the issue of redundancy that is inherent in liver images.

## 3.10 Experimental Results

The comparison of mono modal liver image registration by correlation coefficient and mutual information for different types of liver image are shown in Table 7. The obtained Image registration results using Mutual Information has been presented in Figure 27.

Figure 28a is source ultrasound liver image, Figure 28b is Target ultrasound liver image, Figure 28c performs the rigid body transformation for the source image, i.e., Rotation, Figure 28d shows the matched image.

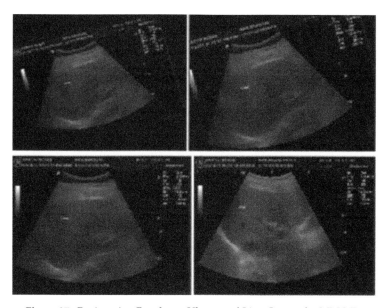

**Figure 27.** Registration Results on Ultrasound Liver Images by MM-MI.

**Table 7.** Performance Analysis of Mono-modal Optimization Techniques.

| Optimization Method | No. of Iterations | Computation Time (Sec) |
|---|---|---|
| DIRECT | 50 | 0.719 |
| NELDE-MEAD | 50 | 0.980 |

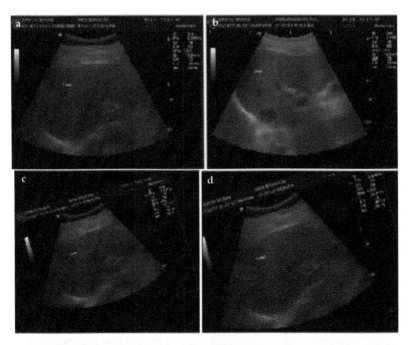

**Figure 28.** Rigid Body Transformation on Mono-modal Ultrasound Liver Images   Translation followed by Rotation. (a) Source image 1, (b) Target image 2, (c) Rotated image, (d) Matched image.

Finally the rigid body transformation is aligned with the reference image correctly placed over the target image. The registered image is then fed into image classification and retrieval to find out similar types of disease images which are present in the database that may be used for further treatment.

# 4

# Texture Feature Extraction

## 4.1 Introduction on Texture Analysis

In addition to color and shape, texture is another prominent and essential feature for the Medical image classification process. The classification activity can be done through similarity matching with the help of the extracted texture features. Texture of an image can be defined by its distinct properties such as coarseness, inherent direction, patter complexity and which are usually sensed from the variations in the scale and orientations in the texture pattern. Texture analysis has become an active research area in the past decade and researchers have proposed numerous methodologies to automatically analyze and recognize texture which helps the medical professionals to carry out the diagnosis. The basic research on texture analysis starts with the derivation of texture energy measures and visual textural features such as coarseness, contrast, directionality, entropy, homogeneity, etc. However most of the existing texture analysis methods were carried out with the assumption that texture images are acquired from the same view point, i.e., same scale and orientation. The multi view representation of the image can enhance the texture analysis process and it has proved to be much more effective in qualifying texture features of a medical image.

Texture is an essential characteristic for the analysis of medical image modalities. Texture contains important information that is used for the interpretation and analysis of medical images. Texture refers to the spatial interrelationships and arrangement of the basic elements of an image. An Ultrasonographic liver image consists of different values of gray-level intensity and different texture. Focal liver lesions are described as regular masses with homogenous internal echoes but diffuse liver lesions are described as masses with a fuzzy border and heterogeneous internal echoes. The Gray Level Co-occurrence Matrices (GLCM) method is a very powerful statistical texture descriptor in medical imaging, especially in Ultrasonic image analysis. One important application of image texture is the recognition of the pathology bearing region using texture properties. Texture classification in medical image processing involves deciding which category a given texture belongs to (Lee et al. 2003). This chapter extends the ideas presented in Chapter 2, considering the fact that the patient's diseases have not been diagnosed and features are not available.

Two important elements for perception of liver disease are different granular sizes and the uniformity of the liver parenchyma. So, we propose a joint method to interpret both the difference of sizes and the gray level between the lobules by auto correlation and, edge-frequency based texture features, respectively.

In this work, the ultrasound images have been taken using a curvilinear transducer array of frequency 4 MHz. These algorithms are applied on 25 samples of normal, and cyst images on 10 * 10 pixel region of interest (ROI) with an 8 bit resolution. The features extracted from these methods are further processed to obtain optimal features which represent the most discriminating pattern space for classification. These optimal features are given as input to the support vector machine for classification purposes.

A feature used to partition images into regions of interest and to classify those regions provides information in the spatial arrangement of colours or intensities in an image characterized by the spatial distribution of intensity levels in a neighbourhood repeating pattern of local variations in image intensity cannot be defined for a point. Texture is a repeating pattern of local variations in image intensity. For example, an image has a 50% black and 50% white distribution of pixels. Three different images with the same intensity distribution, but with different textures.

- Texture consists of texture primitives or texture elements, sometimes called **texels**.
- Texture can be described as fine, coarse, grained, and smooth, etc.
- Such features are found in the tone and structure of a texture.
- Tone is based on pixel intensity properties in the *texel*.
- While structure represents the spatial relationship between **texels**.

- If **texels** are **small** and tonal differences between **texels** are large a fine texture results.
- If **texels** are **large** and consist of several pixels, a coarse texture results.

## 4.2 Texture Analysis

Two primary issues in texture analysis:

- Texture classification
- Texture segmentation

Texture classification is concerned with identifying a given textured region from a given set of texture classes.

- Each of these regions has unique texture characteristics.
- Statistical methods are extensively used (e.g., GLCM, contrast, entropy, homogeneity).

Texture segmentation is concerned with automatically determining the boundaries between various texture regions in an image.

There are three approaches to defining exactly what texture is:

- Structural: texture is a set of primitive texels in some regular or repeated relationship.
- Statistical: texture is a quantitative measure of the arrangement of intensities in a region. This set of measurements is called a feature vector.
- Modeling: texture modeling techniques involve constructing models to specify textures.

Statistical methods are particularly useful when the texture primitives are small, resulting in **micro** textures. When the size of the texture primitive is large, first determine the shape and properties of the basic primitive and the rules which govern the placement of these primitives, forming macro textures.

Texture analysis was first used for satellite image analysis and is increasingly used to evaluate the texture properties in order to improve interpretation of medical images. Image texture of medical images describes internal structure of human tissues or organs as well as pathological changes. Many studies have demonstrated the value of texture analysis in medical applications. The choice of texture analysis rests on its ability to quantitatively describe the image.

Texture analysis has also been applied across various types of imaging modalities such as ultrasound, magnetic resonance imaging (MRI), computer tomography (CT). The employment of texture analysis in medical imaging has been proved to be valuable, particularly for MRI, CT, and ultrasound. Most of the previous works in texture analysis involve MRI images because of the large amount of details provided by this technique. Nevertheless, texture analysis of all sorts of images is possible and has been performed. Texture analysis applications on medical images are summarized according to the type of the imaging modality (ultrasound).

## 4.3 Importance of Dimensionality Reduction

In medical image processing, selecting a minimal set of features based on texture pattern is a special form of dimensionality reduction. Transforming the input image into the set of features is called feature extraction. If the feature extraction is chosen carefully, then it is expected to extract the relevant information from the pathology bearing region (PBR) of the input image in order to perform the desired task, by using this reduced representation instead of the full size image. Features often contain information relative to gray shade, texture, shape or context. In this book, to classify the ultrasound liver diseases present in the image by form of texture, features must be extracted.

Visual criteria for diagnosing focal and diffuse liver diseases are in general confusing and highly subjective. The textural analysis on the ultrasound image includes echogenicity, echotextural variations, liver surface, echogenic properties of hepatic portal vein and so on. Besides the average echogenicity-Gray-Level Co-occurrence Matrix (GLCM), statistical feature matrix and the fractal dimension have also been tried as textural classifiers in medical image analysis (Horng et al. 2002). Diagnostic accuracy using only visual interpretation is currently estimated to be around 72% (Foster et al. 1980).

There are four approaches that are used for texture analysis that are shown in Figure 29. The Structural approaches (Levine 1985) represented texture by well-defined primitives (micro texture) and a hierarchy of spatial arrangements (macro texture) of those primitives. To describe the texture one must define the primitives and the placement rules. The choice of a primitive and the probability of the chosen primitive to be placed at a particular location can be a function of location or the primitives near the location. The advantage of the structural approach is that it provides a suitable symbolic description of the image. However this feature is more useful for synthesis rather than analysis tasks. The powerful tool for structural texture analysis is provided by mathematical morphology (Serra 1982). It may prove to be useful for bone image analysis, e.g., for the detection of changes in bone microstructure.

Diagnostic ultrasound is a useful clinical tool for imaging organs and soft tissues in the human body. Liver disease is widely recognized as an emerging public health crisis. The gray-scale type of display is useful in the detection of tumours. One of its important applications is liver imaging Liver focal diseases are concentrated in a small area of the liver tissues, while the rest of the liver appears normal. In some cases, it may be difficult to diagnose from the image alone, hence a biopsy examination must be conducted.

However, ultrasound automatic diagnosis has a lot of restricted applications in liver diseases identification due to its qualitative, subjective and experience-based nature. The existing methods were all standardized for the scanner to ensure the fidelity of the tissue characterization procedures because the attenuation of ultrasonic waves depends mainly on the emission frequency, and machine settings such as time gain compensation (TCG), tissue enhanced imaging (TEI), focus and gain (G).

Benign lesions contain very fine blood capillaries. The capillary appearance of benign lesions in the image overlaps enough with the portion of distinctive lesions. This makes tissue differentiation very difficult or impossible.

While the liver cyst contour is smooth, the contrast between the normal liver and cyst tissue is high, i.e., the gray level of cyst tissue is much darker than that of normal tissue. After contrast media is injected, the gray level of normal tissue is dramatically enhanced, as is contrast.

A biopsy is the usual method for finding the type of tumours present, but there is a very high risk of post biopsy haemorrhage. It is very difficult, even for experienced physicians, to diagnose the existence, type and the level of a disease in the liver. Therefore, a reliable non-invasive method for early detection and differentiation of these diseases is clearly desirable.

Texture feature refers to the spatial inter relationships of the basic elements of an image. Texture methods can be classified as Statistical, Structural; Model based and Signal Processing techniques. Multi-scale filtering methods such as Gabor wavelet has shown significant potential for texture description, where the advantage is taken of the spatial frequency concept to maximize the simultaneous localization of energy in both spatial and frequency domains. Ahmadian (2003) compared Dyadic and Gabor wavelet and showed that the Gabor wavelet achieves a higher classification rate since it is a rotation-invariant and eliminates the DC components of an input image.

Mona Sharma (2001) compared five different feature extraction methods on Meastex database. The results showed that the Law's method and the co-occurrence matrix method yield a higher recognition rate. Haralick (1979) reviewed various approaches and models for texture. He concluded that for micro textures, the statistical approach works well and for macro texture, histograms of primitive properties and co-occurrence can be used. Kyroacou classified normal, fatty, cirrhoosis and hematoma, ultrasonic images and found that fractal dimension and spatial gray level dependence matrices gave a higher level of accuracy than gray level Run length statistics. Laws and Kenneth Ivan (1980) used Brodatz texture and other images to compare his masks with co-occurrence based features and found that his method had a high success rate. The different types of feature extraction are shown in Figure 29.

## 4.4 Types of Feature Extraction

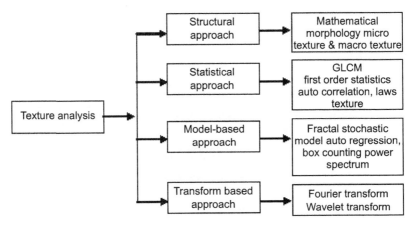

**Figure 29.** Types of Feature Extraction.

## 4.5 Gray Level Co-Occurrence Matrices

A Haralick texture feature based on gray level co-occurrence matrix is used for feature extraction and liver image classification. This Haralick et al. (1973) features captures information about the texture patterns that emerge in the ultrasound liver image. They are calculated by constructing a co-occurrence matrix (Fritz Albregtsen 2008). Once the co-occurrence

100

matrix has been constructed, twelve features can be calculated from its mathematical formula.

One of the defining qualities of texture is the spatial distribution of gray values. The use of statistical features is therefore one of the early methods proposed in the image processing literature. Haralick (1979) suggested the use of co-occurrence matrix or gray level co-occurrence matrix. It considers the relationship between two neighboring pixels, the first pixel is known as a reference and the second is known as a neighbor pixel. In the following, we will use $\{I(x; y); 0 \le x \le N_x-1; 0 \le y \le N_y-1\}$ to denote an image with G gray levels. The G X G gray level co-occurrence matrix $P_x^0$ for a displacement vector $d = (d_x, d_y)$ and direction $\theta$ is defined as follows. The element (i; j) is the number of occurrences of the pair of gray levels i and j which the distance between I and j following direction $\theta$ is d.

$$p_d^\theta (i, j) = \#\{((r, s), (t, v)): I(r, s) = 1, I(t, v) = j\} \tag{26}$$

where

$$(r, s), (t, v) \in N_x \times N_y; (t, v) = (r + dx, s + dy)$$

In addition, there are also co-occurrence matrices for vertical direction ($\theta = 90$) and both diagonal directions ($\theta = 45; 135$). If the direction from bottom to top and from left to right is considered, there will be eight directions (0, 45, 90, 135, 180, 225, 270, and 315). From the co-occurrence matrix, Haralick proposed a number of useful texture features.

Gray-Level Co-occurrence Matrices (GLCM) defined by Haralick, Shanmugan, and Dinstein (1973). GLCM is a matrix that shows the frequency of adjacent pixels with grayscale values i and j. For example, let matrix I be the grayscale values of image I, and (i, j) denotes a possible pair of the horizontally adjacent pixels i and j.

$$I = \begin{bmatrix} 0 & 0 & 1 & 1 \\ 0 & 2 & 1 & 1 \\ 0 & 2 & 2 & 2 \\ 2 & 2 & 1 & 0 \end{bmatrix} \text{where} \quad (i, j) = \begin{bmatrix} (0,0) & (0,1) & (0,2) \\ (1,0) & (1,1) & (1,2) \\ (2,0) & (2,1) & (2,2) \end{bmatrix}$$

GLCM represents the frequency of all possible pairs of adjacent pixel values in the entire image. For instance, in the GLCM for image I (i.e., $GLCM_I$), there is only one occurrence of two adjacent pixel values both being 0 (i.e., (0, 0)), whereas the frequency of having (0, 2) pixel values in image I is two, and so forth.

$$GLCM_I = \begin{bmatrix} 1 & 1 & 2 \\ 1 & 2 & 0 \\ 0 & 2 & 3 \end{bmatrix}$$

From the resulting GLCM, we estimated the probability of having a pair of pixel values (i, j) occurring in each image (i.e., P(i, j)). For example, the probability of having a pair of pixel values (0,0) in image I is 1/12, and the probability of having is Pd,θ. This is a sample of creating GLCM matrix that is shown in the Figure 30.

The 14 features are calculated for all the four types of liver diseases. Those data sets are applied in two forms

- The entire image is considered for the calculation of these metrics
- The infected region will be selected and that particular region is considered for the calculation of 14 Haralick features.

The gray level co-occurrence matrix (GLCM) is a statistical method used for texture analysis. As the name suggests, the co-occurrence matrix is constructed from the image by estimating the pairwise statistics of pixel intensity. A co-occurrence matrix is a two-dimensional array, in which both the rows and the columns represent a set of possible image values. The use of the co-occurrence matrix is based on the hypotheses that the same gray level configuration is repeated in a texture. This pattern will vary more by fine textures than by coarse textures. The co-occurrence matrix $C\alpha,d$ (i, j) counts the co-occurrence of pixels with gray values *i* and *j* at a given distance *d* and in a given direction α. The steps used to construct the co-occurrence matrix are:

- Count all pairs of pixels from the ultrasound liver image in which the first pixel has a value *i*, and its matching pair is displaced from the first pixel by *d* has a value of *j*.
- This count is entered in the *i*th row and *j*th column of the matrix $P_d[i,j]$.

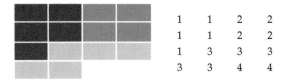

**Figure 30.** (a) Images (b) Matrix.

- Note that $P_d[i,j]$ is not symmetric, since the number of pairs of pixels having gray levels [i,j] does not necessarily equal the number of pixel pairs having gray levels [j,i].

GLCM represents the frequency of all possible pairs of adjacent pixel values in the entire ultrasound liver image. The twelve Haralick's features—Contrast, Correlation, Auto Correlation, Homogeneity, Dissimilarity, Energy, Entropy, Angular Second Momentum, Mean, Variance, Cluster Prominence and Cluster Shade are calculated for liver diseases (Fritz Albregtsen 2008).

Texture analysis algorithms are applied in two stages. First, the entire image is considered for the calculation of these metrics, as shown in the Figure 31. Second the infected region will be selected by the radiologist and that particular region (32 x 32 pixels PBR) is considered for the calculation of Haralick's twelve features by an experienced physician so as to avoid a deviation in image statistics (Pavlopoulos et al. 2000). A number of texture features may be extracted from the GLCM (Haralick et al. 1973, Conners et al. 1984). The following notations are used to describe texture features from the GLCM (Fritz Albregtsen 2008).

- G is the number of gray-levels used.
- μ is the mean value of P.
- $\mu_x$, $\mu_y$, $\sigma_x$, $\sigma_y$ are the means and standard deviations of $P_x$ and $P_y$, that are calculated from the Equation (29) to Equation (32).
- $P_x(i)$ and $P_y(j)$ are the $i^{th}$ and $j^{th}$ entry in the marginal-probability matrix.
- $P_x(i)$ and $P_y(j)$ are obtained by summing up the rows of P(i,j) as given by Equations (27) and (28).

a) $$p_x(i) = \sum_{j=0}^{G-1} p(i,j) \tag{27}$$

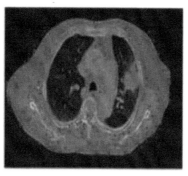

**Figure 31.** (a) Texture Analysis for the Entire Image.

103

b) $p_y(j) = \sum_{i=0}^{G-1} p(i,j)$ (28)

c) $\mu_x = \sum_{i=0}^{G-1} i \sum_{j=0}^{G-1} p(i,j) = \sum_{i=0}^{G-1} i p_x(i)$ (29)

d) $\mu_y = \sum_{i=0}^{G-1} \sum_{j=0}^{G-1} j\, p(i,j) = \sum_{j=0}^{G-1} j p_y(j)$ (30)

e) $\sigma_x^2 = \sum_{i=0}^{G-1} (i-\mu_x)^2 \sum_{j=0}^{G-1} p(i,j) = \sum_{i=0}^{G-1} (p_x(i)-\mu_x(i))^2$

f) (31)

g) $\sigma_y^2 = \sum_{i=0}^{G-1} (i-\mu_y)^2 \sum_{j=0}^{G-1} p(i,j) = \sum_{i=0}^{G-1} (p_y(j)-\mu_y(j))^2$

h) (32)

i) $p_{x+y}(k) = \sum_{i=0}^{G-1} \sum_{j=0}^{G-1} p(i,j) \quad i+j = k$ (33)

For $k = 0, 1\ldots 2(G-1)$.

$$p_{x-y}(k) = \sum_{i=0}^{G-1} \sum_{j=0}^{G-1} p(i,j) \quad |i-j| = k$$ (34)

For $k = 0, 1\ldots G-1$.

## 4.6 Haralick Texture Features

The twelve Haralick's features used in this work are as follows:

### 4.6.1 Contrast

Contrast is a measure of intensity or gray-level variations between the reference pixel and its neighbor. It is used to describe an intensity variation among a homogeneous texture. This measure of contrast or local intensity variation will favour contributions from P (i, j) away from the diagonal, i.e., i ≠ j.

$$\sum_{n=0}^{G-1} n^2 \left\{ \sum_{i=1}^{G} \sum_{j=1}^{G} P(i,j) \right\} \quad |i-j| = n$$ (35)

104

When i and *j* are equal, the cell is on the diagonal and i–j = 0. These values represent pixels entirely similar to their neighbor, so they are given a weight of 0. If *i* and *j* differ by 1, there is a small contrast, and the weight is 1.

### 4.6.2 Correlation

Correlation is a measure of gray level linear dependence between the pixels at the specified positions relative to each other. When correlation is high the image will be more complex than when the correlation is low. GLCM Correlation can be calculated for successively larger window sizes using the Equation 36 (as seen below). Correlation will be high if an image contains a considerable amount of linear structure.

$$\sum_{i=0}^{G-1} \sum_{j=0}^{G-1} \frac{\{i \times j\} \times P(i,j) - \{\mu_x \times \mu_y\}}{\sigma_x \times \sigma_y} \tag{36}$$

### 4.6.3 Autocorrelation

The autocorrelation function of an image can be used to detect repetitive patterns of textures. When coarseness is defined as "composed of relatively large parts or particles", the textural character of an image depends on the coarseness of the texture. An autocorrelation function measures the coarseness of an image; this function evaluates the linear spatial relationships between texture primitives. The autocorrelation function of an image $f(p,q)$ is defined as follows:

$$c_f f(p,q) = \frac{MN}{(M-p)(N-q)} \frac{\sum_{i=1}^{M-p} \sum_{j=1}^{N-q} f(i,j)f(i+p,j+q)}{\sum_{i=1}^{M} \sum_{j=1}^{N} f^2(i,j)} \tag{37}$$

where $(p,q)$ is the positional differences in the $(i,j)$ direction and $M, N$ are the image dimensions. The first part of this definition is to normalize the image boundaries; the second part computes the actual autocorrelation. If the texture evaluated is coarse the autocorrelation function will decrease slowly with increasing distance, because $f(i, j)f(i + p, j + q)$ will often be the same as $f^2(i,j)$. This is because in coarse textures a large number of neighbouring pixels will have the same gray level.

### 4.6.4 Homogeneity

Homogeneity assesses image homogeneousness and for smaller differences between grey values it takes on large values. A homogeneous image will result in a co-occurrence matrix with a combination of high and low P[i,j]'s.

Where the range of gray levels is small the P[i,j] will tend to be clustered around the main diagonal. It is also known as Inverse Difference Moment (IDM).

$$\sum_{i=0}^{G-1} \sum_{j=0}^{G-1} \frac{1}{1+(i-j)^2} P(i,j) \tag{38}$$

IDM is also influenced by the homogeneity of the image. Because of the weighing factor $(1 + (i - j)^2)^{-1}$ IDM will get small contributions from inhomogeneous areas $(i \neq j)$. The IDM value for inhomogeneous images is low, and a relatively higher value for homogeneous images. The IDM has a relatively higher value when the high values of the matrix are near the main diagonal. This is because the squared difference $(i-j)^2$ is then smaller, which increases the value of $1/1+(i-j)^2$.

### 4.6.5 Dissimilarity

The heterogeneity of the gray levels is shown by dissimilarity. Moreover the coarser textures are portrayed by higher values of dissimilarity.

$$\sum_{j=0}^{G-1} P(i,j)^2 \tag{39}$$

### 4.6.6 Angular Second Moment (ASM)

Angular Second Moment evaluates the consistency of textural information. It is also known as uniformity or energy. It measures the uniformity of an image. A homogeneous region will contain only a few gray levels, giving a GLCM with only a few but relatively high values of P(i,j) thus the sum of the squares will be high.

$$ASM = \sum_{i=0}^{G-1} \sum_{j=0}^{G-1} \{P(i.j)\}^2 \tag{40}$$

### 4.6.7 Entropy

The disorderliness of an image is given by entropy. Entropy is a measure of information content. It measures the randomness of intensity distribution when all the elements of the matrix are maximally random; entropy has its highest value. So a homogeneous image has lower entropy than an inhomogenous image. In fact, when energy gets higher, entropy should get lower. Such a matrix corresponds to an image in which there are no preferred gray level pairs for the distance vector d.

$$-\sum_{i=0}^{G-1} \sum_{j=0}^{G-1} P(i,j) \times \log(P(i,j)) \tag{41}$$

Inhomogeneous texture regions like fatty liver may have low first order entropy, while a homogeneous texture region has high entropy.

### 4.6.8 Energy

The energy of a texture describes the uniformity of the texture pattern. In a homogeneous image there are very few dominant gray-tone transitions, hence the co-occurrence matrix of the image will have fewer entries of large magnitude. So the energy of an image is high when the image is homogeneous.

$$Energy - \sum_{i,j=0}^{G-1} \frac{Pd\theta}{(i-j)^2} \tag{42}$$

Energy value can also be computed from Equation (40) and which follows in Equation (43).

$$Energy = \sqrt{ASM} \tag{43}$$

### 4.6.9 Mean

For the symmetrical gray level co-occurrence matrix, each pixel in the image is counted one as a reference pixel and another one as a neighbor pixel. If the two values are identical then the left hand of the equation calculates the mean based on the reference pixel. Otherwise it is also possible to calculate the mean using the neighbour pixels, as shown in the right hand equation.

$$\mu_i = \sum_{i,j=0}^{G-1} i(P_{i,j})\mu_j = \sum_{i,j=0}^{G-1} j(P_{i,j})\mu_i \tag{44}$$

### 4.6.10 Sum of Squares, Variance

Variance is a measure of the dispersion of the values around the mean of combinations of reference and neighbor pixels. It is the sum of the difference between intensity of the central pixel and its neighborhood.

$$\sum_{i=0}^{G-1} \sum_{j=0}^{G-1} (i-\mu)^2 P(i.j) \tag{45}$$

This feature puts a relatively high weight on the elements that differ from the average value of P(i,j).

### 4.6.11 Cluster Prominence

For some of the GLCM texture features, no operations are to be performed on the probability distribution $P(i,j \mid \Delta x, \Delta y)$ that requires the information to be in matrix or histogram form (Peckinpaugh 1991). Using the Cluster shade and cluster prominence Equations (46) and (47), it is possible to reduce the number of addition operations greatly by expressing the texture features in terms of the image pixel values contained in a given neighborhood. This measure indicates that how many clusters of the gray levels present in the image can be classified.

$$\text{Prom} \sum_{i=0}^{G-1} \sum_{j=0}^{G-1} \{i + j - \mu_x - \mu_y\}^4 x\, P(i.j \mid d) \tag{46}$$

**Table 8.** List of Features Extracted from the Liver Image.

| Liver | Energy | ASM | Entropy | Contrast | Mean |
|---|---|---|---|---|---|
| Normal | 1.0899e + 011 | 1.1879e + 022 | 1.4920e + 007 | 2.6962e + 007 | 0.0739 |
| Normal | 1.0787e + 011 | 1.1636e + 022 | 1.5031e + 007 | 2.6945e + 007 | 0.0756 |
| Normal | 1.0311e + 011 | 1.0631e + 022 | 2.3340e + 007 | 2.6872e + 007 | 0.0951 |
| Normal | 1.0058e + 011 | 1.0116e + 022 | 2.4451e + 007 | 2.6833e + 007 | 0.0797 |
| Cyst | 9.9113e + 010 | 9.8235e + 021 | 2.5320e + 007 | 2.6810e + 007 | 0.0820 |
| Cyst | 9.4921e + 010 | 9.0101e + 021 | 2.4335e + 007 | 2.6746e + 007 | 0.0711 |
| Cyst | 1.3047e + 010 | 1.7022e + 020 | 9.9714e + 006 | 3.0330e + 007 | 0.0988 |
| Cyst | 1.6297e + 010 | 2.6558e + 020 | 1.0138e + 007 | 3.0380e + 007 | 0.0978 |
| Cyst | 1.3404e + 010 | 1.7966e + 020 | 9.9962e + 006 | 3.0336e + 007 | 0.0977 |
| Cyst | 3.8542e + 010 | 1.4855e + 021 | 1.0173e + 007 | 2.8978e + 007 | 0.0780 |
| Chirrosis | 3.8050e + 010 | 1.4478e + 021 | 1.3833e + 007 | 2.8970e + 007 | 0.0785 |
| Chirrosis | 3.6098e + 010 | 1.3031e + 021 | 1.4224e + 007 | 2.8940e + 007 | 0.0791 |
| Chirrosis | 1.2813e + 010 | 1.6417e + 020 | 1.5335e + 007 | 3.0244e + 007 | 0.1043 |
| Chirrosis | 1.0575e + 009 | 1.1183e + 018 | 2.3621e + 006 | 3.2407e + 007 | 0.1034 |
| Chirrosis | 1.4296e + 010 | 2.0437e + 020 | 2.4732e + 007 | 3.0267e + 007 | 0.1034 |
| Fatty | 4.4246e + 010 | 1.9577e + 021 | 2.2164e + 007 | 2.9066e + 007 | 0.0824 |
| Fatty | 3.6577e + 010 | 1.3379e + 021 | 2.3275e + 007 | 2.8948e + 007 | 0.0740 |
| Fatty | 1.0087e + 011 | 1.0175e + 022 | 2.4386e + 007 | 2.6838e + 007 | 0.0822 |
| Fatty | 1.0547e + 011 | 1.1125e + 022 | 2.3825e + 007 | 2.6908e + 007 | 0.0852 |
| Fatty | 1.7378e + 011 | 3.0198e+ 022 | 2.4936e + 007 | 2.7959e + 007 | 0.0587 |

### 4.6.12 *Cluster Shade*

In the cluster shade Equation (47), it is possible to reduce the number of operations by expressing this texture feature in terms of the image pixel values contained in a given $M \times N$ neighborhood containing $G$ gray levels from 0 to $G–1$, where f(m,n) is the intensity at sample m, line n of the neighborhood. It calculates the brightness region of the image and groups it, it also identifies the affected region approximately.

$$\sum_{i=0}^{G-1} \sum_{j=0}^{G-1} \{i+j-\mu_x-\mu_y\}^3 x \, P(i.j) \tag{47}$$

Both cluster shade and cluster prominences are measures of the skewness of the matrix, in other words the lack of symmetry. When cluster shade and cluster prominence are high, the image is not symmetric.

## 4.7 Edge Detection

Edge features can be very effective in CBIR when the contour lines of images are evident. The edge feature used in our experiments is the edge detection histogram. To acquire the edge detection histogram, an image is first translated to a gray image, and a canny edge detector is applied to obtain its edge image. Based on the edge images, the edge detection histogram can then be computed. Each edge detection histogram is quantized and is employed to represent the edge feature. The edge detection histogram and corresponding pyramid wavelet transform is shown in the Figure 32.

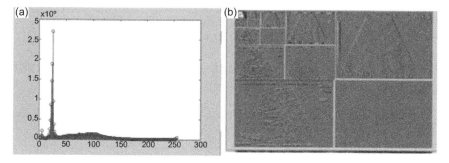

**Figure 32.** (a) Edge Detection Histogram (b) Pyramid Wavelet Transform.

### 4.7.1  Canny Edge Detector

Finds the edge by looking for local maxima of the gradient, it also uses two thresholds to detect strong and weak edges, which are less affected by noise computer to others, and detects true weak edges.

- Gaussian smoothing to select scale.
- Edge detection-differencing convolution or convolve with derivative of Gaussian edge magnitude and direction.
- Suppress non-maxima of derivative (thinning), precise edge locations (sub-pixel precision).
- Track using hysteresis thresholds.

### 4.7.2  Euclidean Distance

Most common use of distance Euclidean distance or simply distance, examines the root of square. This method is simple and relatively faster than the correlation method. The Euclidean distance formula is,

$$E(I, J) = \sqrt{\sum |f_i(I) - f_j(J)|^2}$$

Two important elements for the perception of liver disease are different granular sizes and the uniformity of the liver parenchyma. So, we propose a joint method to interpret both the difference of sizes and gray level between lobules by auto correlation and, edge-frequency based texture feature respectively.

In this work, the ultrasound images have been taken using a curvilinear transducer array of frequency 4 MHz. These algorithms are applied on 25 samples of normal, and cyst images on 10 * 10 pixel region of interest (ROI) with an 8 bit resolution. The features extracted from these methods are further processed to obtain optimal features which represent the most discriminating pattern space for classification. These optimal features are given as an input to the support vector machine for classification.

### 4.7.3  Gabor Texture Parameters

An important desirable feature of the Gabor filter is optimum joint spatial/ spatial-frequency localization, orientation selectivity. The spatial/frequency analysis has played a central role in feature extraction as it combines the two fundamental domains. An important property of the Gabor transform is that its coefficients reveal the localized frequency description of a signal (or) an image, instead of the global frequency information as provided by the coefficients of Fourier Transform. Basically a 2-D Gabor filter is a complex sinusoid ally—modulated Gaussian function of some frequency and orientation.

A 2-D Gabor filter is given by

$$g(x, y) = 1/2\pi\sigma x\sigma y exp [-1/2 [x/\sigma x + y/\sigma y] + \quad 2 \prod j Wx$$

where W is the frequency of sinusoid.

σx and σy characterize the spatial extent and bandwidth of the filter in x and directions respectively. An input image I(x,y) is passed through a set of Gabor filters to obtain the set of filtered output. The Gabor Mean, Standard Deviation, Skewness, and Variance are calculated as texture parameters from the magnitude of the transformed coefficients.

Table 9 shows the optimal Gabor texture parameters. The Gabor parameters depend on the variation of gray levels in pixels. The Gabor mean appears to be less for cyst than benign and normal images. They have less gray level variation and it nearly appears to be homogeneous. The Gabor mean is larger for benign than cyst and normal images. So it is found that the benign image appears to be inhomogeneous. The normal images are intermediate between these two types.

**Table 9.** Gabor Wavelet Features.

| Image Type | Scale Value | Gabor Mean | Gabor Std.Dev | Gabor Variance | Gabor Skew |
|---|---|---|---|---|---|
| | | | Orientation | | |
| Normal | 1 | 1.06 | 0.287 | 0.088 | −0.061 |
| | 2 | 2.08 | 0.129 | 0.016 | −0.012 |
| | 3 | 1.43 | 0.71 | 0.645 | 2.312 |
| Benign | 1 | 1.25 | 0.34 | 0.128 | 2.765 |
| | 2 | 2.99 | 0.16 | 0.022 | 1.744 |
| | 3 | 2.12 | 0.77 | 0.558 | 3.18 |
| Cyst | 1 | 0.35 | 0.01 | 0.453 | −0.03 |
| | 2 | 1.15 | 0.19 | 0.765 | −0.76 |
| | 3 | 0.76 | 0.35 | 0.127 | −0.84 |

## 4.8 Autocorrelation Based Texture Feature

The textural character of an image depends on the spatial size of texture primitives. An autocorrelation function can be evaluated to measure this coarseness. This function evaluates the linear spatial relationship between primitives. If the primitives are large, the function decreases slowly with an increasing distance, whereas it decreases rapidly if the texture consists of small primitives. The set of autocorrelation coefficients shown below are used as texture features and are shown in Table 10.

Where p, q is the positional difference in the i, j direction, and M, N are image dimensions. In this study we vary the (p,q) from (0,0) to (9,9) giving us a total of 100 features.

The table shows that autocorrelation co-efficients decreases slowly with increasing distance, which implies that the images are composed of coarse texture. Since there is a large variation in the gray level values of the pixels in the cyst and benign images, the co-efficient shows low value compared to the co-efficient of the normal images.

**Table 10.** Autocorrelation based Texture Features.

| Feature | Types of Liver Image | | |
|---|---|---|---|
| (P,Q) | Normal | Benign | Cyst |
| 6,4 | 0.84 ± 0.012 | 0.78 ± 0.028 | 0.446 ± 0.089 |
| 6,5 | 0.84 ± 0.012 | 0.78 ± 0.028 | 0.456 ± 0.909 |
| 6,6 | 0.84 ± 0.012 | 0.88 ± 0.032 | 0.56 ± 0.99 |
| 6,7 | 0.84 ± 0.012 | 0.88 ± 0.034 | 0.564 ± 0.10 |
| 8,5 | 0.84 ± 0.014 | 0.88 ± 0.036 | 0.567 ± 0.11 |

## 4.9 Local Binary Pattern

Local Binary Pattern (LBP) is utilized to describe the local texture feature of images. The LBP method provides a robust way for describing pure local binary patterns designed for extracting in a texture feature in ultrasound liver images. The original 3 × 3 neighborhoods threshold is identified by the value of the center pixel. The LBP operator represents the texture in the image by thresholding the neighborhood with the gray value of its center pixel and the results will be described as binary code format. The pixel-to-pixel comparison in the image produces the texture and the resulting image is in the form of a texture histogram. Local Binary Pattern method is a gray scale invariant and can be easily combined with a simple contrast measure by computing for each neighborhood the difference of the average gray level of those pixels which have the value 1 and those which have the value 0 respectively which is shown in Figure 33. The resultant values are summed and then replaced in the pixel value of center.

The mathematical expression for LBP is given below.

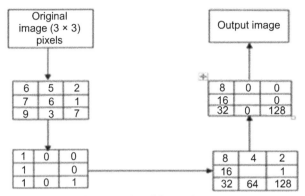

**Figure 33.** Local Binary Pattern.

$$LBP = \sum_{p=1}^{p} 2^{p-1} X f(Gp - Gc)$$

$$f(x) = \begin{cases} 1, x \ge 0 \\ 0, else \end{cases}$$

LBP is a two-valued code. The LBP value is computed by comparing gray value of the center pixel with its neighbors, using the above two equations, where

Gp – gray value of its neighbors

Gc – gray value of the center pixel

P   – number of neighbors

R   – radius of the neighborhood

Liver diseases like hepatoma, cirrhosis, hemangioma can be easily identified based on the texture of the ultrasound liver images.

## 4.10  Edge–Frequency Based Texture Features

A number of edge detectors can be used to yield an edge from an original image. An edge dependent texture description function E is calculated as follows:

$$E(d) = |f(i,j) - f(i+d, j)| + |f(I,j) - f(i-d, j)| + $$
$$|f(i,j) - f(i,j+d)| + |f(i,j) - f(i, j-d)|$$

This function is inversely related to the auto correlation function. Texture features can be evaluated by choosing specified distances, that is

d. Varying the distance d parameters from 1 to 30 giving us a total of 30 features. Here features are based on distance-related gradient. Micro edges can be detected using small distance operators and macro edges need large size edge detectors. The dimensional of texture description is specified by the number of considered distance is represented in Table 11.

The table shows the optimal features that are extracted in which the distances have a small variation to detect the micro edges. The distances are varied to find the gray level differences between a pixel and its neighbouring pixel at various distances. The results show that the gray level differences increase as the distance increases. It implies that the gray level differences are larger for normal images than tumour and cyst images.

Table 11. Edge Frequency Feature.

| Feature D | Types of Liver Image | | |
|---|---|---|---|
| | Normal | Benign | Cyst |
| 1 | 58.4 ± 7.2 | 39.04 ± 8.22 | 6.14 ± 4.59 |
| 2 | 92.6 ± 10.4 | 62.24 ± 16.12 | 11.04 ± 9.02 |
| 8 | 112.4 ± 14.2 | 101.14 ± 21.02 | 14.42 ± 10.2 |
| 14 | 143.7 ± 21.9 | 105.04 ± 21.24 | 15.64 ± 11.62 |
| 21 | 165.5 ± 23.8 | 132.14 ± 32.2 | 15.74 ± 12.21 |

## 4.11 Optimal Feature Sets

The optimal features selected from these methods correspond to a total of 27 features.

Table 12 shows the optimal number of features extracted from each of the methods. They are selected in a way to show a wide variation among all types of images useful in discriminating various types of liver abnormalities. The optimal features are selected from the total number of features obtained from each method (ACF = 99, Gabor = 24, Edge frequency= 30 features).

The performance evaluation of the three texture methods is based on the ability of the classifier to recognize the abnormalities. The texture method that gives the best recognition has high rate of prediction. The motivation of this paper is to find which of the feature extraction methods gives a high predication rate in classifying the abnormalities of ultrasound liver images. Table 13 shows that a high prediction rate is obtained for features extracted from Gabor wavelet followed by Autocorrelation and the Edge texture method.

Table 12. Optimal Features in each Texture Algorithm.

| Feature Extraction Method | No. of Features |
|---|---|
| Gabor Wavelet | 5 |
| Autocorrelation Method | 8 |
| Edge Frequency method | 6 |

Table 13. Prediction Rate of Texture Method.

| Texture Method | Rate of Predication |
|---|---|
| Gabor Wavelet | 77.6% |
| Autocorrelation Method | 68% |
| Edge Frequency method | 64.2% |

From this table it is seen that the Gabor wavelet classifies better when compared to other methods since it analyses the image in both time and frequency domains. But since the ultrasound image is degraded by speckle noise and the appearance of lesion overlaps enough with the sonogram, the result is therefore less in ultrasound liver images.

## 4.12 Scale Invariant Feature Texture

Scale invariant feature transform (SIFT) is an algorithm to detect and describe local features in the ultrasound images. Important characteristics of these features are that the relative positions between them in the ultrasound liver image shouldn't change from one image to another. SIFT can robustly identify objects even among clutter and under partial occlusion, because the SIFT feature descriptor is invariant to uniform scaling, orientation and partially invariant to affine distortion and illumination changes. SIFT involves four different stages for extracting the features from the images and is shown in Figure 34. The stages are scale-space extreme detection, key point localization, orientation assignment and key point descriptor.

$$L(X, Y, \sigma) = G(X, U, \sigma) * I(X, Y)$$

$$G(X, Y, \sigma) = \frac{1}{2\pi\sigma} * e^{-}(x^2 + y^2)/2\sigma$$

$G(X, Y, \sigma)$ is a variable scale Gaussian,

$I(X,Y)$ is the input image,

$\sigma$ is a width of image.

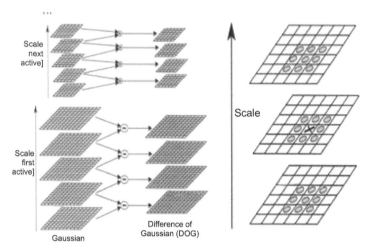

**Figure 34.** Scale Invariant Feature Texture.

SIFT extracted the pathology region of texture features from the ultrasound cirrhosis images. The ultrasound liver image has multiple curves that can be detected from the image and matched to a database of liver images.

## 4.13 Feature Selection

Feature selection plays an important role in classification problems. It can be extremely useful in reducing the dimensionality of the data to be processed by a certain classifier, reducing execution time or even improving predictive accuracy. Diverse feature selection techniques have been proposed in the machine learning literature, such as principal component analysis (PCA), correlation-based feature selection methods, Support vector machine feature elimination, etc.

Kernel Principal Component Analysis (KPCA) is one of the methods available for analyzing ultrasound medical images of liver cancer. First the original ultrasound images need airspace filtering, frequency filtering and morphologic operation to form the characteristic images and these characteristic images are fused into a new characteristic matrix. Then analyzing the matrix by using KPCA and the principle components (in general, they are not unique) and are found in order to the most general characteristics of the original image can be preserved accurately. Finally the eigenvector projection matrix of the original image which is composed of the principle components and can reflect the most essential characteristics of the original images. The simulation experiments were made and effective results were acquired. Compared with the experiments of wavelets, the experiment of KPCA showed that KPCA is more effective than wavelets especially in the application of ultrasound medical images.

The Correlation Feature Selection (CFS) measure evaluates subsets of features on the basis of the following hypothesis: "Good feature subsets contain features highly correlated with the classification" (Hall and Smith 1998). The author suggested the use of the CFS method to extract a minimal set of features that are more suitable for ultrasound liver image detection and classifications. Haralick's defined twelve features based on gray level co-occurrence matrix-Contrast, Correlation, Auto correlation, homogeneity, dissimilarity, cluster shade, cluster prominence, energy, entropy, Angular Second Moment, mean and variance to analyze gray scale texture pattern in ultrasound images.

All its mathematical formula and its importance over medical datasets are discussed above. CFS defined in Equation (48) is applied over GLCM. From the statistical measure that is analyzed in the Table 4.2, correlation feature selection has selected most dominant five features namely Contrast, Cluster prominence, Auto correlation, Cluster shade and Angular Second Moment (ASM) out of the twelve Haralick's features. Correlation feature selection method is used to reduce the space complexity for medical image classification process.

$$Merit_{s_k} = \frac{k\overline{r_c f}}{\sqrt{k + k(k-1)\overline{r_{ff}}}}. \tag{48}$$

where Merit Sk is the correlation between the summed features and the disease class and 'k' is the number of features. Here, $\overline{rcf}$ is the average of the correlations between the features and class, and $\overline{rff}$ is the average inter-correlation between features.

## 4.14 Metrics

The proposed techniques are used for assessing the focal and diffuse liver diseases by analyzing the characteristics of liver echo-texture. It not only generates quantitative measures to assess liver disease progression but also classifies the disease stages of ultrasonic liver images. In clinical diagnosis, the amount of texture in the pathology bearing region (PBR) is one of the key factors used to assess the progression of liver diseases. Five of the twelve features (contrast, cluster prominence, cluster shade,

Angular second momentum, and Auto correlation) used are found to be the strongest describers by CFS of a texture which can be observed from feature extraction results. Texture properties coupled with fractal produced a good accuracy rate. Experimental results show that the analysis of the obtained results suggested that diseases like Cyst, Fatty Liver, and Cirrhosis can be diagnosed with only five features namely Contrast, Auto Correlation, Angular Second Momentum, Cluster prominence and Cluster Shade out of twelve features belonging to Haralick's textural features. Besides, fractal —a model based approach is used to distinguish diseased livers from the normal livers.

## 4.15  Case Study

Texture is an important significant property of medical images based on which images can be characterized and classified in a Content Based Image Retrieval and Classification system. This paper examines the feature extraction methods to ameliorate texture recognition accuracy by extracting the rotation invariant texture feature from liver images by individual Gabor Filter method and by the MGRLBP (multi-scale Gabor rotation-invariant LBP) method. The features extracted from both the approaches are tested on a set of 60 liver images of four different classes. The classification algorithms such as Support Vector Machine (SVM) and K-Nearest Neighbor (KNN) were used to evaluate the extracted features from both methods, showing advancing improvements with the MGRLBP method over the individual method in the classification task.

## Related Works

Visual features of liver tissues can be explained numerically in a number of ways (Amores and Radeva 2005). The simple gray-level distributions are frequently used, because the intensities describe physical properties of liver tissues effectively. Second-order statistics such as the gray-level co-occurrence matrices (GLCM) and run length (RLE) (Schlomo et al. 2006) have also been widely incorporated for additional feature information. Another type of popular feature extraction technique is based on filters, which are also used for certain applications in liver imaging.

Ludvik Tesar et al. (Brown et al. 1997) examined the specific 3D extent of Haralick features for effective segmentation of three-dimensional CT scans of the abdominal area. This approach is not suitable for complex textures. V. Subbiah Bharathi and L. Ganesan (Brown 1992) used a new statistical approach for the discrimination of normal liver and abnormal liver using orthogonal moments as features due to the extraneous nature but their computation time is high. Yali Huang et al. (Yali Huang et al. 2010) extracted texture features from the ultrasonic liver images using three different spatial techniques such as Gray Level Histogram, Gray Level Difference Statistics and Gray Level Co-occurrence Matrix which significantly discriminates the normal liver and fatty liver images. The size of dataset used is too small in this approach. Farhan Riaz Francisco Baldaque Silva proposed a novel descriptor, autocorrelation Gabor features (AGF) for the classification of Gastroenterology images using invariant Gabor descriptors. Farhan used autocorrelation homogeneous texture (AHT) as a region based descriptor but they did not use color descriptors for classification.

Deepti Mittal et al. proposed a system, that uses four different feature extraction techniques such as First Order Statistics (FOS), Spatial Gray Level Dependency Matrix (SGLDM), Gray Level Run Length Matrix (GLRLM) and Texture Energy Measures (TEM) to identify focal liver lesions in B-mode ultrasound images but the classification accuracy is low. Ranjan Parekh used Gray Level Co-Occurrence Matrix and Wavelet Decomposition matrix methods for feature extraction process to recognize human skin diseases. The author considered only 8 bit images for analysis. U. Rajendra Acharya et al. proposed a descriptive method for Glaucoma detection using a combination of Texture and Higher Order Spectra Features (HOS) in which the Feature extraction was performed using GLCM and Run Length Matrix Methods and the classification was performed with SVM, naive Bayesian and random-forest classifiers. The limitation with this approach was that the size of the diverse medical images used for classification were too small.

Yang Song et al. proposed an automatic classification method for different categories of HRCT lung images in which higher descriptiveness was obtained by combining multiple methods such as Local Binary Pattern, Gabor filters and Histogram of Oriented Gradients method. The limitation with this approach was that the classification accuracy of the complex lung tissue category was low. Omer Kayaalti et al. proposed a non-invasive and fast approach to specify fibrosis using texture properties. Omer used Grey Level Co-Occurrence Matrix, Discrete Wavelet Transform and Discrete Fourier Transforms for feature extraction. The success rate of this approach is very low and not sufficient for clinical diagnosis. Jitendra Virmani et al. differentiated normal and cirrhotic liver based on Laws' Masks Analysis. Jitendra used a methodology of designing a computer aided diagnostic system with optimal Laws' texture features and a neural network classifier for effective discrimination between normal and cirrhotic segmented regions but the dataset used for the classification was too small.

## Research Contribution

Medical image classification system is a system which classifies the multi-categories of medical images from the available large digital medical database, through feature extraction for encoding the image features as feature descriptors. This feature extraction process extracts the most promising and effective feature as descriptors from the query images. In the proposed system a rotation-invariant texture feature descriptor is extracted using individual rotation-invariant Gabor Filter method and by MGRLBP method, which combines the rotation invariant property of LBP and multi-scale property of Gabor filters. Based on the extracted rich texture feature, the liver images are classified using Machine learning algorithms such as Support Vector Machine (SVM) classifier and k-Nearest neighbor classifier. The performance of the texture classification can be measured by calculating the accuracy of the classifiers.

## Proposed MGRLBP Method

A texture feature is a characteristic that can capture a certain visual property of an image either globally for the whole image, or locally for objects or regions. Texture is the variation of data at different scales and at different rotations. The proposed MGRLBP method incorporates two feature extraction methods such as Local Binary Pattern (LBP) and Gabor Filter to extract the rotation invariant texture feature which discriminates the normal liver from the affected liver effectively. The LBP feature describes the spatial structure of local image texture in which the rotation invariance can be achieved. On the other hand, the multi-scale and multi-orientation representation of Gabor filters is often demonstrated as a highly effective texture descriptor and its multi-scale nature is quite useful for computing multi-resolution LBP features. Therefore, to incorporate rich texture information while attempting to minimize intra-category variations, the proposed MGRLBP feature extraction method combines the multi-scale property of Gabor filters and the rotation-invariant property of LBP features. The proposed system is shown in the Figure 35.

The rotation invariance texture feature is extracted in the system using Gabor filters in which all Gabor functions of certain scale with different orientation angles as 30°, 45°, 60° and 90° are summed together. Then for each Gabor-filtered image, a rotation-invariant LBP feature is computed for each pixel in the image as

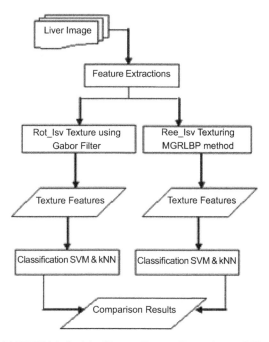

**Figure 35.** MGRLBP Method for Texture Feature Extraction and Classification.

$$I^{s}(x,y) = \sum_{r=0}^{R-1} I^{s,r}(x,y) \tag{49}$$

where r and s denotes the orientations and scale and $I^{s,r}$ denotes the Gabor filtered images, obtained by convolving image I with Gabor functions of R orientations. On the other hand frequency and orientation representations of Gabor filters resemble the human visual system, and they have been found to be very effective for representing texture. In the spatial domain, a 2D Gabor filter is a Gaussian kernel function modulated by a sinusoidal plane wave. The rotation invariant texture feature can be extracted by modulating the conventional Gabor filter method with respect to a constant scale, i.e., by summing all the orientations of 30°, 45°, 60° and 90° at each scale level, which extracts features from a specific scale band covering all the orientations of the specified liver image.

## Dataset Evaluation and Metrics

For the evaluation purpose, ultrasound liver images of four different categories such as Hemangioma, Fatty Liver, Cyst Liver and Normal Liver are collected from various hospitals and Scan centers along with the guidance of the medical professionals. There are in total 56 images of which 15 images are normal, 15 images are Hemangioma, 13 images are Fatty liver, and 13 images are Cyst liver.

To evaluate the proposed method, invariant texture features are extracted from all the four different categories of liver tissues using both the conventional Gabor filter and also by the MGRLBP method. The rich invariant texture is extracted under different orientations as $30°$, $45°$, $60°$ and $90°$ with constant scale, describes the unique variations in each tissue category. Then the medical image classification task is performed based on the extracted invariant texture features. Two standard classifiers such as Support Vector Machine (SVM) and k-Nearest Neighbor (kNN) were used. Support Vector Machine is supervised learning models. It takes a set of input data, and checks which of the two classes predicts the output, making it a non-probabilistic binary linear classifier. More formally, a support vector machine constructs a hyperplane or set of hyperplanes in a high- or infinite-dimensional space, used for classification, and the hyperplane that has the largest distance to the nearest training data point of any class reduces the generalization error of the classifier. The distance between the image features is calculated using the Euclidean distance formula,

$$E(I,J) = \sqrt{\sum |f_i(I) - f_j(J)|^2} \tag{50}$$

where $f_i(I)$ and $f_j(J)$ denotes the feature values of images I and J. Similarly kNN is a non-parametric learning algorithm. k-NN assumes that the data is in a feature space, and operates on the premises that classification can be done by relating the unknown to the known based on certain distance or similarity function. Each set of the training data consists of a set of vectors and class label associated with each vector. It may be either + or – (for positive or negative classes). But k-NN, can work equally well with arbitrary number of classes.

## Results and Discussions

Performance of the classification is measured by calculating retrieval accuracy using the following relation

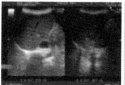

**Figure 36.** Implementation of MGRLBP method on Hemangioma Liver Category.

**Table 14.** Classification Accuracy of Texture Features for Various Liver Tissue Categories.

| Liver Tissue Categories | MGRLBP | Gabor |
|---|---|---|
| Hemangioma | 82.52 | 77.87 |
| Fatty Liver | 80.70 | 71.26 |
| Cyst Liver | 79.88 | 73.34 |
| Normal Liver | 88.31 | 79.65 |

$$\text{Accuracy} = \frac{\text{True Positives} + \text{True Negatives}}{\text{True Positive} + \text{True Negatives} + \text{False Positives} + \text{False Negatives}}$$

where true positive refers to the proportion of the positive images that are correctly identified and true negative refers to the proportion of the negative images that are correctly identified from the total positive and negative images. Performance of the proposed MGRLBP feature extraction method is compared with the conventional Gabor method. MGRLBP method gives a more satisfied classification results (88% accuracy) than the individual feature extraction method (79% accuracy). Figure 36 shows the implementation of MGRLBP method on Hemangioma liver category in which the Gabor filter with multiple orientations are summed together. Then the Local Binary Pattern is applied for the Gabor filtered image and the changes in intensities are observed with the histogram. The classification accuracy of texture features for different liver tissues is shown in Table 14 and Table 15 shows the performance comparison between the feature extraction methods. From Table 15, it can be inferred that performance of

**Table1 15.** Performance Comparison between Feature Extraction Methods.

| S. No. | Classification Techniques | Gabor Filters | MGRLBP Method |
|---|---|---|---|
| 1 | k-Nearest Neighbor | 73.33 | 86.65 |
| 2 | Support Vector Machine | 79.86 | 88.23 |

the classification process is enhanced with the compound method rather than the conventional method.

## 4.16 Conclusion & Future Work

A combined feature extraction method is proposed which increases the texture discriminative ability on different classes of liver images and it reduces the intra class variations within the medical image. The accuracy of the classification system is also improved to (88%) in MGRLBP method when compared with the individual feature extraction methods (79%). In future, texture feature can be combined with other features such as gradient, intensity, etc. to calculate the feature vector and also the classification is performed on other classification schemes such as Sparse based approximation method and Dictionary Construction method.

# 5

# Image Classification and Retrieval

## 5.1 Introduction

Image classification and retrieval plays a vital role in the medical domain. A minor error in disease identification can result in increased financial costs in diagnosis and other consequences in treatment. In such cases, an automated system that could classify and retrieve medical images based on user's interest and thus provide decision support is an essential requirement. This type of mechanization could be a semi-automated or fully-automated process. The learning mechanism has to be properly developed since an improper design will result in large number of misclassifications during liver imaging.

The earlier chapters in this book have presented the directions and conclusions in the areas of preprocessing, registration, and feature extraction. However, the robustness of the results could be improved by using appropriate approaches that present automated decision making and as such has been considered as the major focus in this chapter. This chapter deals with applying various machine learning techniques for classification and retrieval of ultrasound liver images.

Content-based image retrieval (CBIR) technology has been proposed to benefit not only the management of increasingly large image collections, but also to aid clinical care, biomedical research, and education. Based on a literature review, we conclude that there is widespread enthusiasm for CBIR in the engineering research community, but the application of this technology to solve practical medical problems is a goal yet to be realized. Furthermore, we highlight "gaps" between desired CBIR system functionality and what has been achieved to date, present for illustration a comparative analysis of four state-of-the-art CBIR implementations using the gap approach, and offer suggestion to overcome the high-priority gaps that lie in CBIR interfaces and functionality that better serve the clinical and biomedical research communities.

Content-based image retrieval (CBIR) technology exploits the visual content in image data. We propose to benefit the management of increasingly large biomedical image collections as well as to aid clinical medicine, research, and education. We treat CBIR as a set of methods that (1) index images based on the characteristics of their visual content, and (2) retrieve images by similarity to such characteristics, as expressed in queries submitted to the CBIR system. These characteristics, also referred to as "signature", may include intensity, color, texture, shape, size, location, or a combination of these. Sketching a cartoon, selecting an example image, or a combination of both methods, is typically used to form the query. The retrieved results are usually rank-ordered by some criteria; however, other methods, such as clustering of similar images, have been used to organize the results as well.

The practical application of CBIR depends on many different techniques and technologies applied at several stages in the indexing and retrieval workflow, such as: image segmentation and feature extraction; feature indexing and database methods; image similarity computation methods; pattern recognition and machine learning methods; image compression and networking for image storage and transmission; Internet technologies (such as JavaScript, PHP, AJAX, Applet/Servlet); and human factors as well as usability. More recently, natural language processing has also been included, in attempts to exploit text descriptions of image content and the availability of standardized vocabularies (Schlomo et al. 2006). It is through careful selection of appropriate methods from these fields that a successful CBIR application can be developed.

The technical literature regularly reports on experimental implementations of CBIR algorithms and prototype systems, yet the application of CBIR technology for either biomedical research or routine clinical use appears to be very limited. While there is widespread enthusiasm for CBIR in the engineering research community, the incorporation of this technology to solve practical medical problems is a goal yet to be realized. Possible obstacles to the use of CBIR in medicine include:

- The lack of productive collaborations between medical and engineering experts, which is strongly related to the usability and performance characteristics of CBIR systems

- The lack of effective representation of medical content by low-level mathematical features
- The lack of thorough evaluation of CBIR system performance and its benefit in health care
- The absence of appropriate tools for medical experts to experiment with a CBIR application, which is again related to the usability and performance attributes of CBIR systems.

Therefore, we take these four factors: content, features, performance, and usability as foundational in classifying and comparing CBIR systems, and in this chapter we use these concepts as (1) an organizational principle for understanding the "gaps", or what is lacking in medical CBIR systems, (2) a lens for interpreting the main trends and themes in CBIR research over the past several years, and (3) a template for a systematic comparison of four existing biomedical CBIR systems.

The concept of gaps has often been used in CBIR literature, with the semantic gap being the most prominent example, we have treated this "concept of gaps" as a paradigm for a broad understanding of what is lacking in CBIR systems and have extended the gap idea to apply to other aspects of CBIR systems, beyond the semantic gap. We may consider the semantic gap to be a *break* or discontinuity in the aspect of image understanding, with "human understanding" on one side of the gap and "machine understanding" on the other. Similarly, we may identify breaks or discontinuities in other aspects of CBIR systems, including the level of automation of feature extraction, with full automation on one side and completely manual extraction on the other; and, for another example, the degree to which the system helps the user refine and improve query results, with "intelligent" query refinement algorithms based on user identification of "good" and "bad" results on one side, and no refinement capability at all on the other. Each gap (1) corresponds to an aspect of a CBIR system that is explicitly or implicitly addressed during implementation; (2) divides that aspect between what is potentially a fuller or more powerful implementation from a less powerful one; and (3) has associated with it methods to bridge or reduce the gap.

## 5.2 Introduction on Machine Learning Techniques

Machine learning algorithm plays an important role in the medical imaging field, including in computer-aided diagnosis, image segmentation, image registration, image-guided therapy and image classification. Recently, the application of machine learning techniques to medical image retrieval has received more attention (Lehmann et al. 2005). Due to the rapid development of computer technology, it is becoming more and more convenient to acquire, digitally store and transfer medical imagery. Nowadays, many hospitals need to manage several tera-bytes of medical image data each year (Muller et al. 2004). Therefore, categorization of medical images is becoming imperative for a variety of medical systems, especially in the application of digital radiology such as CAD and Case-based reasoning (Lehmann et al. 2005). This chapter deals with different machine learning algorithms namely—Support Vector Machine, Neural Network, Relevance feedback and Fuzzy logic classifier for classification and retrieval of ultrasound liver images.

## 5.3 Supervised Vs Unsupervised Medical Image Classification

### 5.3.1 Supervised Learning Algorithm

The Support Vector Machine (SVM), stemming from statistical learning theory, involves state-of-the-art machine learning. The SVM based on statistical learning theory is widely used in supervised learning and classification problems, invariably providing a good performance in medical image classification problems (Lee et al. 2003). Compared to the conventional neural network, the SVM has the advantage of being usable under different kernel functions and highly accurate classification based on parameter selection. The kernel function $k(x, y)$ is used as a mapping or a transformation function. The separation of data can be either linear or non-linear. The different types of kernel function are linear, nonlinear, polynomial, RBF, and quadratic. The SVM classifier is a part of computer aided diagnosis, which assists radiologists in accurately diagnose liver diseases (Chien-Cheng Lee et al. 2007).

### 5.3.2 Unsupervised Learning Algorithm

The Self Organizing Map (SOM) is an unsupervised learning algorithm where the input patterns are freely distributed over the output node matrix (Kohonen 1990). SOM utilizes the concept of competitive learning for medical image classification. Competitive learning is an adaptive process in which the neurons gradually become sensitive to different input categories. The SOM is a type of artificial neural network that produces a low-dimensional discretized representation of the input space of the training samples called a map. SOM uses a neighborhood function to preserve the topological properties of the input space. Klein Gebbinck et al. (1993) applied discriminate analysis with a neural network to differentiate between various types of diffuse liver diseases.

### 5.3.3 Relevance Feedback Classifier

In the past years, relevance feedback techniques have evolved from early heuristic weighting adjustment techniques to various machine learning techniques for medical image classification and retrieval (Huang and Zhou 2001). Relevance Feedback (RF) is a query modification technique, originating in information retrieval that attempts to capture the user's precise needs through iterative feedback and query refinement. Most research in relevance feedback uses one or both of the following approaches: (1) query-point moving and (2) weight updating. The query-point moving approach tries to improve the estimation of the ideal query point by moving the current query point (i.e., estimate) by a certain amount based on user feedback. The weight updating approach is a refinement method based on modifying the weights or parameters used in the computation of similarity based on the user's feedback.

A Support Vector Machine with Relevance feedback system for the classification and retrieval of ultrasound liver images is implemented for a small dataset (150 liver images). The hybrid approach (HSVM+RF) comprises several benefits when compared to existing Content Based Image Retrieval (CBIR) for medical system. In general in the CBIR system, image registration is not necessary for retrieval of similar images but in medical image retrieval, image registration plays an important role in monitoring the growth of the liver pathologies. It reduces the redundancy of retrieval images from the database. This work is coupled with both gray level co-occurrence matrices (GLCM) and Fractal texture feature to extract features of the PBRs. Through the selection of significant features by Correlation based Feature Selection (CFS), the input spaces can be simplified, which then is forwarded to the SVM as input. In order to reduce the semantic gap during image classification and retrieval, the future iteration will be preceded by the physician selection of appropriate images from the retrieved images by applying the relevance feedback technique.

Thus the classification and retrieval for ultrasound liver images is completed by HSVM-RF. The performance of this machine learning method has been tested with a 150 liver dataset. The result had an accuracy of 72.1%. With this HSVM method, the classification and retrieval of ultrasound liver images is done only for two categories (i.e., Positive samples or Negative samples). It could be understood from the results that multiclass SVM with selected five Haralick's GLCM features gives an accuracy rate of 81.7%. The choice of using multiclass SVM with Haralick's feature vs. HSVM with RF depends on the computational complexity of the approach and multiclass levels. For the large dataset with different stages of liver diseases, the classification is switched to unsupervised learning algorithm.

### 5.3.4 Fuzzy Classifier

Fuzzy classifier plays an important role in dealing with uncertainty decision making in medical applications (Kuncheva Ludmila and Fried Rich Steimann 1999). Classification of medical image objects is based on association of a given object with one of several classes (i.e., diagnoses). Fuzzy logic is a form of mathematical logic in which truth can take continuous values between 0 and 1. Jesus Chamorro et al. (2010) showed that the propositions of the fuzzy logic can be represented with degrees of truthfulness and falsehood. A Membership Function (MF) is a curve that defines how each point in the input space is mapped to a membership value (or degree of membership) between 0 and 1. The input space is sometimes referred to as the universe of discourse. Jin Zhao and Bimal (2002) discussed that the membership function is similar to the indicator function in classical sets.

From the above discussion, Computer-based classification and retrieval of medical image data using machine learning algorithm can simplify the difficult interpretation of medical images and provide additional information to the physician. In order to be appropriate in clinical practice, it has to cope with the following two requirements:

1. During the training phase, untypical cases (due to imaging errors, abnormal physiological parameters of the patients, or untypical symptoms) have to be considered with a lower weighing factor so that the knowledge acquired during the training contains the typical characteristics of the classes.
2. The result of classification of unknown cases has to contain information on how reliable the computer-based diagnosis is.

Both requirements are not met by conventional classifiers. This work concentrates more on the accuracy of classification and retrieval of ultrasound liver images for automated decision making. The motivation behind this research is aimed at image classification according to the nature of pathology bearing region (PBR) that appears across a given image dataset containing liver images. The proposed work concentrates on classifying liver diseases like Cyst, Fatty Liver, Hepatoma, Hemangioma, and Cirrhosis from ultrasound Normal images which is shown in Table 16.

**Table 16.** Ultrasound Liver Dataset used for Classification and Retrieval.

| US Liver | Description | Liver Images |
|----------|-------------|--------------|
| Cyst | Typically round, anechoic, smoothly delineated structures with refraction shadows at the edges of the liver. Cysts are thin-walled structure that contains fluid. Most cysts are single, although some patients may have several cysts called Polycystic Liver Diseases (PLD). | |
| Hepatoma | Irregular-shaped often vague contours, the complex structure of vessels and the heterogeneity of the tissue. Hepatocellular carcinoma (HCC, also called malignant hepatoma) is the most common type of liver cancer. | |
| Hemangioma | Uniformly well circumscribed hyperechoic mass less than 3 cm in diameter. These lesions are often small and can be multiple. Female hormones may promote the formation and growth of hemangioma. | |
| Fatty liver | Hypoehoic spared areas, The lack of a mass effect and tapered margins are part of the usual appearance of sparing. Fatty liver, also known as fatty liver disease (FLD) is a reversible condition, where large vacuoles of triglyceride fat accumulate in liver cells via the process of steatosis. | |
| Cirrhosis | The surface nodularity is shown clearly, facilitated by the presence of ascites. A liver normally contains a certain amount of fat due to alcohol intake, but if fat represents over 5–10% of the weight of the liver, that person is said to have cirrhosis. | |

## 5.4 Literature Review

Yang et al. (1988) developed a scoring system to evaluate the severity of liver diseases based on several image characteristics such as echogenicity, attenuation and masking of the portal vein or gall bladder. Steatosis can also be classified on the basis of image features. Liver diseases are best identified using these gray scale images. The most useful tissue differentiation techniques are based on the investigation of B-scan images, as presented by Momenan et al. (1990). Ultrasound B-scan images present various granular patterns as texture. Hence the analysis of an ultrasonic image will lead to the problem of texture classification. Texture is an image feature that provides important characteristics for surface and object identification from image (Tuceryan and Jain 1993).

Rolland et al. (1995) suggested that the problem of searching medical images is a very important one that deserves more attention particularly due to the significant false diagnosis reported by clinics. Pavlopoulos et al. (1996) evaluated the performance of different texture features for a final quantitative characterization of ultrasonic liver images. Bleck et al. (1996) used the autoregressive periodic random field model to classify normal and fatty liver. Yasser Kadah et al. (1996) extracted first order gray level parameters like mean and first percentile and second order gray level parameters like contrast, Angular Second Moment, Entropy and Correlation and trained the functional link neural network for automatic diagnosis of diffuse liver diseases like fatty and cirrhosis using ultrasonic images and showed that very good diagnostic rates can be obtained using an unconventional classifier trained on actual patient data.

E-Liang Chen et al. (1998) used Modified Probabilistic Neural Network on CT abdominal images in conjunction with feature descriptors generated by fractal feature information and the gray level co-occurrence matrix and classified liver tumors into hepatoma and hemangioma with an accuracy of 83%. There are a large number of other important techniques to improve the performance of classification and retrieval systems. One of the most prominent techniques is relevance feedback, because often unwanted images show up in the result of a classification. The active selection of relevant and irrelevant images by the doctors represents an interactive method for getting more accuracy of retrieval images (Rui et al. 1998). Hong et al. (2000) incorporated SVM to Content Based Image Retrieval (CBIR) with Relevance feedback. The author proposed the importance of relevance feedback in CBIR. Wen-Li et al. (2003) proposed a feature selection algorithm based on fractal geometry and M-band wavelet transform for the classification of normal, cirrhosis and hepatoma ultrasonic liver images. Amores and Radeva (2005) presented a CBIR system for intravascular ultrasound images using a generalization of correlogram in order to extract local, global and contextual image features.

Chien-cheng Lee et al. (2007) proposed the multiclass SVM classifiers to distinguish CT liver images-cyst, hepatoma and hemangioma based on GLCM features. Based on its results, the similarity in the features of liver cyst and hemangioma makes the liver disease classification difficult. Several scientists have investigated the liver tissues classification by the features derived from ultrasonic images. Nicolas et al. (2010) first utilized texture features to classify livers and spleens of normal humans. The proposed work concentrates on improving the learning capabilities of the model considering the fuzzy nature of the dataset.

### 5.4.1 SPIRS

The Spine Pathology and Image Retrieval System (SPIRS) was developed at the U.S. National Library of Medicine to retrieve x-ray images from a large dataset of 17,000 digitized radiographs of the spine and associated text records. Users can search these images by providing a sketch of the vertebral outline or selecting an example vertebral image and some relevant text parameters. Pertinent pathology on the image/sketch can be annotated and weighted to indicate importance. This hybrid text-image query yields images containing similar vertebrae along with relevant fields from associated text records, which allows users to examine the pathologies of vertebral abnormalities.

### 5.4.2 Interface

SPIRS provides a Web-based interface for image retrieval using the morphological shape of the vertebral body. A query editor enables users to pose queries either by sketching a unique shape, or by selecting or modifying an existing shape from the database. Additional text fields enable users to supplement visual queries with other relevant data (e.g., anthropometric data, quantitative imaging parameters, patient demographics). These hybrid text-image queries may be annotated with pertinent pathologies by selecting and weighting local features to indicate importance. Query results appear in a customizable window that displays the top matching results and related patient data.

Significant gaps that are yet to be addressed in the SPIRS system are similar to those for CervigramFinder, and include, for Feature Gaps, lack of multiscale analysis; for Performance Gaps, lack of integration into use in a biomedical system and lack of quantitative evaluation; for Usability Gaps, no user query refinement. (However, see comments about "data exploration" below.) Capabilities that have at least partially addressed some gaps include, for Content Gaps, manual labeling of vertebrae by anatomical type; for Feature Gaps, computer-assisted feature extraction (an Active Contours algorithm is used to find approximate boundaries of vertebrae in the images; these boundaries then are manually reviewed and corrected); for Performance Gaps, feature vector indexing by K-D Tree, and qualitative evaluation; and, for Usability Gaps, support for both query by composition (see Section 2.1.1) and by interactive user sketch. We also note that SPIRS provides capability to specify not only the shape to be used in the query, but *which part of the shape should be used*, so that the user may focus on the fine level of structure that is often critical in biomedical image interpretation. In addition, SPIRS provides (1) "basic" user feedback on each returned image, namely, a figure of dissimilarity to the query image; and (2) a "data exploration" capability, which takes query results as a beginning point to initiate new and related queries; using a given query result, that is, a vertebral shape returned by a query, the entire spine containing that shape may be displayed; then the user may select a vertebra in that same spine and use its shape as a new query. It should be noted that SPIRS, like CervigramFinder, *operates on local,* region-of-interest data in the image. The system characteristics of SPIRS indicate that it is for research, teaching, and learning on 2D data; it accepts as input, and creates output "hybrid" data (both text and image). In this regard, SPIRS allows the user to specify as a query a vertebral shape and some text (such as age, race, gender, presence/ absence of back or neck pain, and vertebra tags such as "C5", to indicate the class of vertebrae being searched for). It then returns such text, along with the associated image data.

### 5.4.3 IRMA

The Image Retrieval in Medical Applications (IRMA) has the following goals: (1) automated classification of radiographs based on global features with respect to imaging modality, body orientation with respect to the x-ray beam (e.g., "anterior-posterior" or "sagittal"), anatomical body region examined, and the biological system under investigation; and (2) identification of local image features including their constellation within a scene, which are relevant for medical diagnosis. These local features are derived from priori classified and registered images that have been segmented automatically into a multi-scale approach. IRMA analyzes content of medical images using a six-layer information model: (1) raw data, (2) registered data, (3) feature, (4) scheme, (5) object, and (6) knowledge.

The IRMA system that is currently available via the Internet retrieves images similar to a query image with respect to a selected set of features.

These features can, for example, be based on the visual similarity of certain image structures. Currently, the image data consists of radiographs. It uses a reference database of 10,000 images categorized by image modality, orientation, body region, and biological system. The system architecture has three main components: (1) the central database, containing images, processing schemes, features, and administrative information about the IRMA workstation cluster; (2) the scheduler, which balances the computational workload across the cluster; and (3) the Web server, which provides the graphical user interface to the IRMA system for data entry and retrieval.

### 5.4.4 Gaps and System Characteristics

In contrast to the rather general concept within the IRMA project, the IRMA system that is currently demonstrated on the web has some significant gaps that are still yet to be addressed. These gaps include, for Content Gaps, lack of semantic labeling; for Feature Gaps, only operation in global image characteristics is supported, and multiscale analysis is lacking; for Performance Gaps, lack of integration into use in a biomedical system, lack of feature vector indexing, and lack of quantitative evaluation. Capabilities that at least partially address system gaps include, for Feature Gaps, fully automatic feature extraction (facilitated, of course, by the fact that IRMA operates on the image as a whole, so that segmentation of particular regions-of-interest prior to feature extraction is not required); for Performance Gaps, a widely-publicized and mature online Internet presence, and qualitative retrieval evaluation; and, for Usability Gaps, an extremely flexible query refinement mechanism that lets the user step back and forth among queries done in a session, and lets the user combine queries with union, intersection, and negation operators. This is coupled with an advanced feedback measure that assists the user in judging how closely a retrieved image matches not only a single image used in the query, but how closely it matches a weighted set of images. The system characteristics of IRMA indicate that it is for research, teaching, and learning use on 2D data.

### 5.4.5 SPIRS-IRMA

IRMA, described above, aims at providing visually rich image management through CBIR techniques applied to medical images using intensity distribution and texture measures taken globally over the entire image. This approach permits queries on a heterogeneous image collection and helps identify images that are similar with respect to global features, e.g., all chest x-rays in the AP (anterior-posterior) view. However, the IRMA system lacks the ability to find particular pathology that may be localized in specific regions within the image. In contrast, the SPIRS system provides localized vertebral shape-based CBIR methods for pathologically sensitive retrieval of digitized spine x-rays and associated metadata. In the SPIRS system, the images in the collection must be homogeneous, i.e., a single modality imaging the same anatomy in the same view, e.g., vertebral pathology expressed in spine x-ray images in the sagittal plane.

Recent work in fuzzy-based medical image classification schemes has focused on extracting the fuzzy features. A fuzzy dominant texture descriptor is proposed for semantically describing an image (Jesus et al. 2010). These approaches do not take advantages of fuzzy set theory in the classification stage. Juan Miguel et al. (2012) presented Fuzzy based medical image retrieval for X-ray images. The existing system makes use of Cobb angle as a feature for fuzzy based image retrieval. The angle helps in the diagnosis of the severity of the disease and this angle is measured manually; therefore there is deviation in the measurements. Hence in order to predict the extent of damage in any modalities, it is essential to eliminate this imprecision. Elimination of this imprecision can be done by employing the concept of fuzzy logic. The approach of fuzzy logic for this application can eliminate imprecision and can produce accurate results for the submitted query image. The accuracy helps the physician or the doctor in proper and accurate diagnosis thereby enhancing the treatment of the patient.

## 5.5 Proposed Methodology

Most of the existing approaches use artificial neural network classifiers. But the accuracy of retrieving similar images consists of classifying more irrelevant images. The process of using machine learning is generally flexible and therefore a lot of decisions are taken in an ad-hoc manner. The objective of this research is to develop an efficient classification and retrieval system for classifying ultrasound liver diseases (both focal and diffuse liver) by three machine learning techniques namely Hybrid Support Vector Machine with Relevance Feedback, Hybrid Kohonen SOM and Fuzzy classifiers. The research concentrates on improving the learning capabilities of the model considering the fuzzy nature of the dataset. From the study, it can be understood that the retrieval plays a vital role and has lot of challenges in research perspective. However in this work, the retrieval is considered only from retrieval of liver images from image database without confronting the techniques being used for retrieval.

### 5.5.1 *Hybrid Support Vector Machine with Relevance Feedback (HSVM-RF)*

The proposed work on HSVM-RF is used to address the main issue—'semantic gap' in image classification and retrieval. The objective of the HSVM-RF method is proposed to classify whether an ultrasound liver image is normal or a diseased one. The systematic flow diagram for the HSVM-RF system is shown in Figure 37. An input query image, i.e., liver image is fed into the proposed hybrid system. After image registration, feature extraction and feature selection, it is compared with liver images in the database pool and the resultant images will be retrieved. The results obtained with the Support vector machine are further improved by applying the relevance feedback technique. Image registration combined with SVM-RF classification makes the system hybrid.

Support Vector Machine is a supervised learning technique used for ultrasound liver image classification and retrieval. For supervised learning, a set of training data and category labels are available and the classifier is designed by exploiting this prior known information. One of the important problems in medical imaging is two-class classification. The binary SVM classifier takes a set of input data and predicts each given input, in which two of the possible classes the data belong to. The original data in a finite dimensional space is mapped into a higher dimension space to make the separation easier. The vectors classified nearer to the hyper plane is called support vectors. The distance between the support vector and the hyper plane is called the margin; the higher the marginal value gives lower the error of the classifier.

SVM is originally a method for binary classification (Lee et al. 2002), however, in medical practice, the number of possible disease types is rarely restricted to two categories, called positive samples and negative samples. Relevance Feedback (RF) in other hand is incorporated in this system to improve the performance of SVM classification by getting feedback from the physician. RF is used in retrieval system for reducing semantic gap (i.e., a gap between high level concepts perceived by the user and low level features that are used in the system).

In this research work the linear kernel function for SVM based classification is utilized. Linear kernel function for binary classification of medical images $K(K-1)/2$ models are constructed, where k is the number of categories. In Linear classification $y = \text{sign}(w.x + \omega_0)$ and its decision surface hyper plane is defined by $w.x + \omega_0 = 0$. Samples on the hyper plane are called the support vectors. The most common pair-wise classification: $K(K-1)/2$ binary classifiers are applied on each pair of classes. Each liver sample is assigned to the class getting the highest number of ranks. A rank

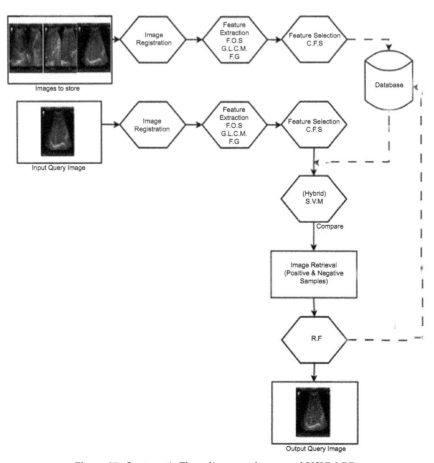

**Figure 37.** Systematic Flow diagram of proposed HSVM-RF.

for a given class is defined as a classifier assigning the pattern to that class. The pair-wise classification is more suitable for ultrasound liver image classification problems. Therefore, the pair-wise approach is adopted in this work. The systematic flow diagram of proposed HSVM-RF is shown in Figure 37.

The procedure of Hybrid SVM with Relevance Feedback is as follows:

1) The Pathology Bearing Region (PBR) is extracted from the input query image by physician or radiologists and registered into the database.
2) Mutual information based image registration on ultrasound liver mono-modal images is done only for registering the patient details

138

whose details are already stored in the database. Mutual information is an automatic, intensity-based metric, which does not require the definition of landmarks or features such as surfaces and can be applied retrospect. MM-MI based image registration has been discussed in Chapter 3 and is used for growth monitoring of liver diseases. Otherwise, PBRs from all query images are directly stored into the database for classification and retrieval.

3) Feature extraction is a crucial step for any pattern recognition task especially for ultrasonic liver tissue classification, since liver images are highly complex and it is difficult to define a reliable and robust feature vector. It is done by First Order Statistics, Fractals and Gray level co-occurrence matrices (GLCM). However, textural features are those characteristics such as smoothness, fitness and coarseness of certain pattern associated with the image. The detailed information of feature extraction and feature selection has been discussed in Chapter 4.

4) Feature Selection is done by Correlation based Feature Selection (CFS). It is known that certain features are appropriate for the classification of a specific disease and other specific features are suitable for other diseases. Therefore, the CFS algorithm (Kwang et al. 2002, Hsu et al. 2002), a feature subset selection technique, is used in this research work which is already discussed in Chapter 4. The goal of feature subset selection is to identify and select the most influential variables from a large pool of variables and further reduce the feature space for classification.

5) Appropriately selected features are then fed to the HSVM classifier with linear kernel function. HSVM classifies both positive (normal liver) and negative (diseased liver) samples based upon the input query. Diagram for SVM classification with linear kernel function is shown in Figure 38.

6) The future iteration will be preceded by the physician to select appropriate images from the retrieved images by applying relevance feedback technique.

Relevance Feedback on another hand is a supervised active learning technique used to improve the effectiveness of classification and retrieval systems. Relevance feedback algorithms proposed deal with a two-class problem, i.e., relevant or irrelevant images. The main idea is to use positive and negative samples from the user to improve system performance. For a given query, the system first retrieves a list of ranked images according to Euclidean distance metrics. Then, the physician marks the retrieved

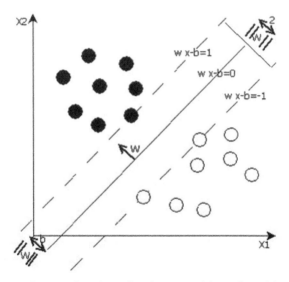

**Figure 38.** Diagram for HSVM classification with linear kernel function.

images as relevant (positive samples) to the query or not relevant (negative samples).

When only positive samples from the physician feedback or when the physician considers only the relevant images, various schemes can be applied, making use of the image content to improve retrieval accuracy. In this specific context, the physician's suggestion can easily be interpreted by means of feedback, add the newly obtained positive samples into the query set, and return the algorithm to retrieve results. In this way, the vector y will have multiple non-zero components that will spread their ranking scores in the propagation process. And the sequence f (t) converges to

$$f = \beta(1 - \alpha S)^{-1} y_i \beta(1 - \alpha S)^{-1} \sum_{i=1}^{n^*+1} \tag{51}$$

where $\alpha$, $\beta$ are suitable constants, $y_i$ is an n-dimensional vector with the $i^{th}$ component equal to 1 and others equal to 0, and n is the number of positive samples fed back by the physician. '$f$' is the one that moves towards positive samples. Therefore these samples will spread ranking scores independently, and assign large value to images belonging to their corresponding neighborhood; the ultimate ranking score is the sum of these individual scores.

Due to the asymmetry between relevant and irrelevant images, the system should process images differently. For example, in the Rocchio formula (Gletsos et al. 2003), the initial query is moved towards positive samples and away from negative samples by different degrees. Edward Chang and Beitao Li (2003) proposed the MEGA system with RF method. In this system positive samples are used to learn the target concept in k-CNF. While negative samples are used to learn a k-DNF that bounds the uncertain region; some researchers have even come up with the idea of introducing different penalizing factors for positive and negative samples into the optimization problem of SVM. A deeper reason for this asymmetry is that relevant images tend to form certain clusters in the feature space, while irrelevant images occupy the remaining feature space.

To accommodate this asymmetry, in the hybrid approach, both positive and negative samples spread their ranking scores differently by using the Equation 52 and Equation 53. To speak precisely, the work first defines two vectors Y+ and Y−. The element of the former one is set to 1 if the corresponding image is the query or a positive sample; while the element of the latter one is set to −1 if the corresponding image is a negative sample. All the other elements of the two vectors is set to 0. Generally, positive samples should make more contribution to the final ranking score than negative samples. Secondly, the work modifies the neighborhood of a negative sample by changing the iteration value. It also controls the neighborhood size within which the points will have a big similarity value to the center point. The formula to propagating negative ranking scores is:

$$f = \beta(y_i^- + \alpha S y_i^{-1} + \alpha S(\alpha S y_i^-) + \dots \tag{52}$$

$$f^* = \beta \sum_{i=0}^{M} \alpha^i S^i y_i^- \tag{53}$$

Thus the neighbourhood of negative samples is smaller than that of positive samples and the scope of their effect is decreased. The experiment is evaluated using performance metrics sensitivity, specificity and retrieval accuracy rate. If the query image is relevant with the database image then it is considered as a positive feedback sample. If query image is irrelevant with the database image then it is considered as a negative feedback sample. If the user is satisfied with the result, then the retrieval process is ended. If the user is not satisfied with the result, relevance feedback plays its role again. Finally, the most relevant images are retrieved.

The system will refine the retrieval results based on the feedback and present the new list of images to the user. Hence, the key issue in relevance feedback is how to incorporate positive and negative samples to refine the query and/or to adjust the similarity measure. Methods for performing relevance feedback using the visual features as well as the associated keywords (Semantics) in unified frameworks have been performed by reducing the semantic gap.

## 5.6 Results and Discussion

### *5.6.1 Hybrid Support Vector Machine with Relevance Feedback*

Initially this work was carried out with 150 images of liver dataset; 40 liver cysts, 20 hepatoma, 40 fatty liver, 25 alcoholic cirrhosis and 25 normal images. The input query image of size 512 x 512 pixels were saved at 12 bits per pixel and the pathology bearing region of gray level was fed into the proposed hybrid system based on its needs. If the patient wants to know about the history of his/her diseases, register his new image in to the database through MM-MI based image registration method. Otherwise, simply store its PBR of input query image into database for classification. Then from the registered images, several important features have been collected namely—gray level co-occurrence matrices, First order statistics and fractal geometry which are more relevant features for ultrasound modality images. It is known that certain features are appropriate for the classification of specific disease and other specific features are suitable for other diseases. Therefore, CFS algorithm, a feature subset selection technique, was used in this work. The goal of the feature subset selection was to identify and select the most influential variables from a large pool of variables and further reduces the feature space for classification. After image registration, feature extraction and feature selection, it is stored into the database and compared with the liver images in the database pool and the resultant images are retrieved.

Figure 39 shows the resultant of retrieved images from the database by HSVM-RF for an input query image. The proposed HSVM-RF concludes

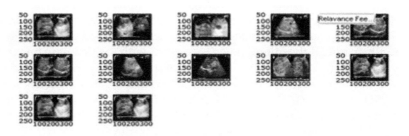

**Figure 39.** Retrieved liver images by HSVM with Relevance Feedback.

that the given input image belongs to either disease image (cyst, cirrhosis, hepatoma or fatty) or normal image. The future iteration will be preceded by the physician to select appropriate images from the retrieved images by applying relevance feedback technique. The result of the relevance feedback is categorized in two categories—positive samples and negative samples which shown in Figure 40 and Figure 41.

Figure 40 shows that the positive samples which are selected by physician or radiologist at each rounds by relevance feedback technique.

Figure 41 shows that the negative samples which indicate the query image which is not relevant to the set of images stored in the database. In HSVM-RF, only two classes can be classified. So this work is extended to Multiclass SVM.

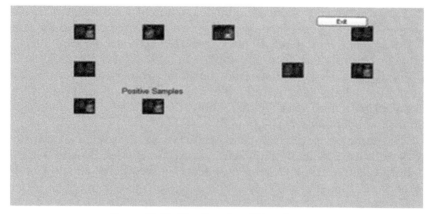

**Figure 40.** Positive Samples of Liver Images.

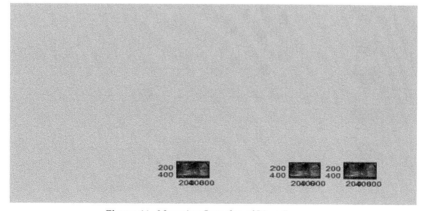

**Figure 41.** Negative Samples of Liver Images.

### 5.6.2 *Multiclass SVM Classifier*

SVM is originally a method for binary classification. However, in medical practice, the number of possible disease types is rarely restricted to two classes. When a multi class classification scheme is conducted, the first step is to divide all classes into binary sub problems. Then, use the SVM method to classify these sub-problems again. For decomposing the multi class problem, one-vs.-all decomposition strategy has been used in this work. The textural features from the input query is extracted by using selected five GLCM features namely contrast, auto correlation, cluster shade, cluster prominence and angular second moment which are then fed into the multi class SVM classifier. To construct the multi class classification based on the one-vs.-all strategy, three binary classifiers, named C1, C2, and C3, with 150 liver dataset are combined to form a hierarchical multiclass classifier scheme for liver disease distinction. This work is carried out with the same 150 images of liver dataset; 40 liver cysts, 20 hepatoma, 40 fatty liver, 25 alcoholic cirrhosis and 25 normal images. In the first layer, the classifier C1 distinguishes the normal tissue from the others. The second layer classifier differentiates focal liver from the other abnormal tissues. Diffuse liver is identified in the third layer. Figure 42 shows the multiclass SVM in nonlinear ultrasound liver dataset.

The performance of HSVM and Multiclass SVM classifier is evaluated by Sensitivity, Specificity and Retrieval Accuracy rate. Sensitivity is to identify positive results for the test set. The sensitivity of a test is the proportion of images which have the malignant tumor in which its test

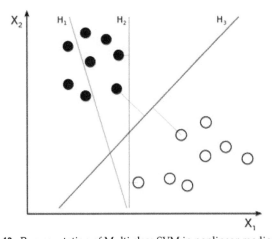

**Figure 42.** Representation of Multi class SVM in nonlinear medical data.

result is positive for it. If a test sample has high sensitivity, a negative result would suggest the absence of disease.

$$Sensitivity = \frac{True\ Positives}{True\ Positive + False\ Negative} \qquad (54)$$

Specificity is to identify negative results for the test set. The specificity of a test is defined as the proportion of images which does not have the malignant tumors. These images will be targeted as negative. If a test has high specificity, then it has high probability of the presence of malignant tumor.

$$Sensitivity = \frac{True\ Negatives}{True\ Negatives + False\ Positives} \qquad (55)$$

The Retrieval Accuracy rate is calculated by the following equation

$$Accuracy = \frac{True\ Positives + True\ Negatives}{True\ Positive + True\ Negatives + False\ Positives + False\ Negatives}$$

$$(56)$$

The retrieved images may belong to positive sample or negative sample. In order to reduce False Negative Rate (FNR), and False Positive Rate (FPR), RF is incorporated into the SVM to obtain very good classification results. The description of both FNR and FPR is shown in the Table 17.

From the above experiments it is concluded that the HSVM-RF with GLCM, FOS, and Fractal features retrieved resultant images for a given query image with an accuracy of 72.1% for a small dataset containing a 150 liver dataset. Whereas the results from the multi class SVM classifiers with five GLCM features (Contrast, Auto Correlation, Angular Second Moment, Cluster Prominence and Cluster Shade) discussed in Chapter 4 show the classification accuracy rate is 81.7% which is comparatively better than the feature extraction by other methods for the same dataset which was our earlier work. The comparative results are shown in Table 18. The key findings from this work shows that diagnosis of liver images can be completed with minimal features which help to reduce overall retrieval time.

**Table 17.** Parameter Description of FNR and FPR.

| Parameters | Description |
|---|---|
| False negative Rate (FNR) | FNR is the probability that the classification result indicate a normal liver while the true diagnosis is indeed liver diseases, i.e., positive. |
| False Positive Rate (FPR) | FPR is the probability that the classification result indicates liver diseases while the diagnosis is indeed a normal liver. |

The result of the proposed method—HSVM-RF is coupled with image registration and is discussed below. Image Registration is an extended part of Chapter 3 considering the fact that the patient liver images have already been diagnosed and recorded as history using Mutual Information. The performance evaluation of HSVM with Registration by MI is shown in the Table 19.

From the Table 19, the mutual information value of a normal liver is about 0.5145 and a fat deposit liver is about 1.4340 nearer to the MI of Fatty liver. This new image is fed into HSVM to check whether similar types of disease image are available in DB. If so, it retrieves similar category of liver images and helps the physician to monitor the growth of the liver diseases. And also, the retrieval accuracy rate for the classification with image registration is about 86% and reduces the chances for redundancy of retrieval images. However, the performance of the proposed work-SVM with relevance feedback, together with image registration is poor due to the following reason. Unlike registration for diagnosis and treatment in the other organs, accuracy requirements are reduced for applications in the liver. Liver tumor is usually large, ranging in diameter from 1.2 to 18.8 cm in clinical trials (1). The liver is very resilient, and often a tumor can be diagnosed with little danger of serious complications. This is fortunate, since motion and deformation in the liver likely degrade the accuracy of rigid body registration. Even though image registration in classification and retrieval reduces the redundancy of irrelevant images and helps to monitor the growth of liver disease for further treatment.

**Table 18.** Comparison of classification accuracy rate for supervised learning methods.

| Comparison | Feature Extraction Methods | Accuracy Rate |
|---|---|---|
| HSVM-RF | GLCM, First Order Statistics, Fractal | 72.1% |
| SVM with multiclass | Selected Five GLCM & Fractal textural features | 81.7% |

**Table 19.** Performance Evaluation of HSVM with Registration by Sensitivity, Specificity and Retrieval Accuracy Rate.

| Liver images | Count | Mutual Information | Sensitivity | Specificity | Retrieval Accuracy rate |
|---|---|---|---|---|---|
| Normal | 25 | 0.5145 | 92 | 88 | 90 |
| Cyst | 40 | 0.5658 | 92.5 | 80 | 86 |
| Fatty | 40 | 1.5003 | 95 | 87.5 | 91 |
| Hepatoma | 20 | 1.9598 | 90 | 90 | 90 |
| Cirrhosis | 25 | 1.5322 | 72 | 80 | 76 |

The Classification and Retrieval point of view investigated from the Table 20 is discussed below. Though SVM with RF works well, when the number of positive feedback samples is small, the performance of SVM with RF becomes poor. This is mainly due to the following reasons:

1. An SVM classifier is unstable for a small-sized training set, i.e., the optimal hyperplane of SVM is sensitive to the training samples when the size of the training set is small. In this hybrid method, the optimal hyperplane is determined by the feedback samples. The users would only label a few images and cannot label each feedback (registered) sample accurately all the time. Therefore, the performance of the system may be poor with insufficient and inexactly labeled samples.
2. SVM's optimal hyperplane may be biased when the positive feedback samples are much less than the negative feedback samples. In the relevance feedback process, there are usually many more negative feedback samples than positive ones. Because of the imbalance of the training sample for the two classes, SVM's optimal hyperplane will be biased toward the negative feedback samples. Consequently, SVM with RF may mistake many query irrelevant images as relevant ones.
3. Finally, in relevance feedback, the size of the training set is much smaller than the number of dimensions in the feature vector and, thus may cause an over fitting problem. Because of the existence of noise, some features can only discriminate the positive and negative feedback samples but cannot distinguish the relevant or irrelevant images in the database. So, the learned SVM classifier, which is based on the feedback samples, cannot work well for the remaining images in the database.

Due to these several issues prevailing in SVM with the RF algorithm, this work is extended to an unsupervised learning algorithm—Self Organizing Map. Performance of the Proposed SVM-RFs compared with SVM. Proposed SVM-RFs give more satisfied retrieval results than SVM which is shown in Figure 43 and Figure 44.

**Table 20.** Retrieval Accuracy Rate for Ultrasound Liver Images.

| Liver Images | Count | Sensitivity | Specificity | Retrieval Accuracy Rate |
|---|---|---|---|---|
| Normal | 25 | 92 | 88 | 90 |
| Cyst | 40 | 92.5 | 80 | 86 |
| Fatty | 40 | 95 | 87.5 | 91 |
| Hepatoma | 20 | 90 | 90 | 90 |
| Cirrhosis | 25 | 72 | 80 | 76 |

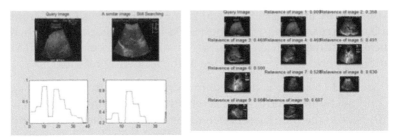

**Figure 43.** (a) Searching similar image using SVM (b) Image retrieval using SVM.

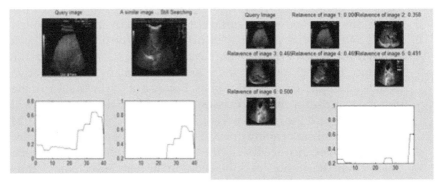

**Figure 44.** (a) Searching similar image using SVM-RF (b) Image retrieval using SVM-RF.

## 5.7 Importance of Medical Image Retrieval

A revolution in the field of medical science has evolved that has led to the betterment in the field of diagnosis. The implementation of image retrieval systems for medical images will eventually assist the physician in better diagnosis of the corresponding disease. The complete manual diagnosis of the medical images by the physician may not be accurate sometimes resulting in erroneous predictions which will have a major effect on the treatment provided to the patient and such mispredictions will have adverse effects the health of the individual. On the other hand, it is hard for the physician to scan through all the archived images manually to diagnose the disease. So to make the work simple and efficient it is advisable to develop a medical image retrieval system to assist in the task of diagnosis.

Among the various applications of computer science in the field of medicine, the processing and analysis of medical image data is playing an increasingly significant role in the present situation. With various colored and grey scale medical imaging techniques such as X-Ray, computer tomography, magnetic resonance imaging, and ultrasound, the amount of digital images that are produced in hospitals is increasing incredibly.

148

Images are ubiquitous in biomedicine and the image viewers play a central role in many aspects of modern health care. Tremendous amounts of medical image data are captured and recorded in digital format during the daily clinical practice, medical research, and education (in 2009, over 117,000 images per day were captured in the Geneva radiology department alone). Facing such an unprecedented volume of image data with heterogeneous image modalities has led to the necessitiy to develop an effective and efficient medical image retrieval system for clinical practice and research. Thus the need for systems that can provide efficient retrieval of images of particular interest that could assist the physician in further treatment is becoming very high.

## 5.8 Types of Image Retrieval

There are various approaches for medical image retrieval such as Text Based Image Retrieval (TBIR) and the Content Based Image Retrieval (CBIR). The former was a conventional approach which was replaced by the latter due to some of the shortcomings of the conventional approach. So far many medical image retrieval systems have been developed using either of these two approaches.

### 5.8.1 Text Based Image Retrieval

This approach was introduced in the late 1970s. In text based image retrieval, every image requires a document representation and annotation that holds precise information about the image which is textually described with the help of captions and stored in a database. This forms the image dataset. The query image which is also textually described is compared against the dataset. Similar text descriptors are matched and the results are retrieved. In a text-based search engine, the system aims to retrieve information that is relevant (or similar) to the user's query. In document retrieval, the query is usually a word or phrase.

Text based medical image retrieval (TBIR) is fast and reliable. But it does not produce exact search results because it requires a well annotated image database. Also there can be many ways to represent the same annotation (Synonymity). In addition to the above shortcomings, common errors such as spelling mistakes, all annotations may not be complete, certain information are hard to express and improper abbreviations are prominent. Thus, Text based image retrieval is a conventional approach with many notable defects. In order to overcome these drawbacks and to improve the field of image retrieval in medical routine, we go for the modern method that would prompt us with exact and precise results that would act better than the text based image retrieval systems.

### 5.8.2 Content Based Image Retrieval

Content based image retrieval refers to the recall of images from a database that are relevant to a query submitted, using information derived from the images themselves, rather than relying on accompanying text indices or other annotation. CBIR has received increasing attention as a result of the availability of large image databases in medicine, science, commerce, and the military. CBIR has been proposed to overcome the difficulties encountered in textual annotation for large image databases. Like a text-based search engine, a CBIR system aims to retrieve information that is relevant (or similar) to the user's query. In CBIR, the querying is done using an image.

The key to successful CBIR lies in the development of appropriate similarity metrics for ranking the relevance to the query image of images in a database. In CBIR, quantitative image features, computed automatically, are used to characterize image content. The image features may be extracted at either a low level or at a high level or both. The query image is then compared to the images in the database on the basis of the measured features. Those images in the database having the highest similarity to the query image are retrieved and displayed for the user.

Some of the image features include color, texture, shape, etc. Most medical images are grayscale and so color features are not commonly used in medical image retrieval. Texture refers to the spatial arrangement of the pixels or intensities in an image. In the medical field, texture features serve as an important tool for diagnosis because they reflect details within an image structure. Shape features include visual characteristics such as curves, surfaces, contour and so on. In medical images only a few pixels bear pathological information. Hence it is mandatory to extract the appropriate features that would assist the diagnosis of pathological regions in the image. This would guide the medical expert to carry out the diagnosis with minimal effort.

In content based image retrieval systems the queries, used to retrieve images, can be classified as primitive, logical and abstract. Primitive query is based on features, such as color, shape and texture, extracted from images. Logical query employs the identities of the objects in the image. Sketch-based and linguistic queries in which the user describes objects or regions in the desired spatial positions and ascribes attributes, such as class label, size, color and shape properties, to them can also be considered as logical queries. Abstract queries are based on notion of similarities. Logical and abstract queries are sometimes called semantic queries.

Although Content Based Image Retrieval serves as a great tool for diagnosis, at times visually similar images may not refer to the same pathology. For instance, images of tumor may visually appear the same but one might be malignant and the other need not be. In general, high feature similarity may not always correspond to semantic similarity. Hence, a system that combines the above two approaches effectively is currently being formulated.

## 5.9  Relevance Feedback using Soft Label SVM

We first describe the ways to log the feedback histories of users and how to employ them to help the regular relevance feedback tasks in CBMIR systems. In a typical relevance feedback procedure, a CBMIR system first returns N images from databases to a user based on some kinds of similarity measure. The user then specifies which images are relevant (positive) and which ones are irrelevant (negative). When the physicians submits his/her judgment results to the CBMIR system, such a feedback session including N+ positive samples and N-negative samples will be logged in the log database. The task for the log-based relevance feedback is to employ the accumulated feedback sessions in the log databases for improving relevance feedback in the retrieval tasks.

The first step toward the log-based relevance feedback is to access and effectively organize the feedback information of users from the log database. For solving this problem, a Relevance Matrix (RM) is constructed to express the relevance relationship between the image samples in the database. The column of the relevance matrix represents the ID of patient image samples in the database and the row represents the session number of feedback session in the log database, the relevance values of positive samples are recorded as relevant (+1), irrelevant (–1_, or unknown (0). For example, suppose image i is marked as relevant and j is marked as irrelevant in a given session k, then the corresponding valued in the matrix is RM (k; i) = +1 and RM(k; j)= –1[7]. Therefore, relationship of two images j and i can be computed by the following.

Modified correlation formula:

$$R_{ij} = \sum \Phi_k . RM(l;i) . RM(k;j)$$

$$\Phi_k = \{ \begin{array}{l} 1 \text{ if } RM(k;i) + RM(k;j) \geq 0, \\ 0 \text{ if } RM(k;i) + Rm(k;j) < 0, \end{array}$$

Where $R_{ij}$ represents the relevance relationship and the $-k$ term is engaged to remove the element pair $(-1; -1)$ for the correlation formula. If $R_{ij}$ is positive, it indicates that image I and image j are relevant otherwise they are irrelevant. Then for each given image sample, we can find a set of relevant samples and a set of irrelevant samples ranking by their relationship. Hence, by engaging the log information, we can obtain a list of relevant samples and irrelevant samples associated with different relationship values in a relevance feedback procedure. The challenge of the log-based relevance feedback is how to formulate effective algorithms to deal with the training samples associated with different confidence degrees in relevance or irrelevance. In order to solve this problem, we propose a modified SVM technique called Soft Label Support Vector Machine with soft labels of different confidence degrees (Carrillo et al. 2000).

### 5.9.1 Soft Label Support Vector Machine

Regular SVM techniques assume that the labels of the data are absolutely correct. However, when some labels of data are incorrect, they may hardly impact on the predicting decision (Fookes et al. 2012, Maes et al. 2003). In order to reduce the errors that arose from the unconfident data, we incorporated the label confidence degree of data in the regular SVM model and proposed the Soft Label SVM as below. Assume the labels of the data are with different confidence degrees.

From this objective function, we can see that the margin error $\xi_i$ will be large if constrained with a small label value $s_i$ and will be smaller for a large one in non-separable cases. This indicates that the Support Vector with the large label will have a larger impact on the decision boundary than the SV with a smaller label. Hence the probability of misclassification on the future data will be reduced.

## A. Algorithm

1) Calculate relevance scores $f_R(Z)$ for all image samples.
2) Choose training samples with Soft Labels based on their relevance scores.
3) Train a Soft Label SVM classifier on the selected training samples with Soft Labels ($f_{SLSVM}(z)$).
4) Rank images based on the combination of the two relevance functions $f_R(z)$ and $f_{SLSVM}(z)$.

## B. Algorithm LRF-SLSVM

**Input:**
Q ← a query sample by a user
L ← set of labeled training samples

**Variables:**

| | |
|---|---|
| S | ← set of "soft label" training samples |
| $C_H$, $C_S$ | ← regularization parameter |
| c | ← correlation or relationship between two images |
| $R_p$, $R_n$, $f_R$ | ← Log based relevance degrees to the query |
| ▲ | ← selection threshold for "soft label" samples |
| $f_{SLSVM}$ | ← a soft label SVM classifier |
| $f_q$ | ← the overall relevance function |

**Output:**

$R_{top}$ ← set of most relevant samples

Init : $R_p(i)$, $R_n(i)$ ← $-\alpha$

BEGIN
1. Compute log based relevance feedback
   for each positive z ∈ L {
      for each $z_i$ ∈ Z {
      $c(i)$ ← comprelationship($z$, $z_i$);}
      c ← Normalize©
      $R_{p(i)}$ ← max($R_{p(i)}$, $c(i)$; }
   for each negative z ∈ L {
      for each $z_i$ ∈ Z {
      $c(i)$ ← comprelationship($z$, $z_i$);}
      c ← Normalize©
      $R_{n(i)}$ ← max($R_{n(i)}$, $c(i)$; }

2. Select "soft label" training samples for each $z_i$ ∈ Z {
   If $R_{p(i)} - R_{n(i)} \geq$ ▲
   then S ← S U { $z_i$ }; }

3. Train a soft label SVM classifier

   $f_{SLSVM} \leftarrow$ Train_soft_label_
   SVM(L, S, $C_H$, $C_s$)

   $f_{SLSVM} \leftarrow$ Normalize ($f_{SLSVM}$)

4. Rand images based on $f_{SLSVM}$

   And $(R_p - R_n)$

   $f_R \leftarrow$ normalize $(R_p - R_n)$

   $f_q \leftarrow f_{SLSVM} + f_R$

   $R_{top} \leftarrow$ Sort_indescending_
   order ( $f_q$ )

   return $R_{top}$

END

## 5.10  Active Learning Methods

Contrary to passive learning, in which the learner randomly selects some unlabelled images and asks the user to provide their labels, active learning selects images according to some principle, hoping to speed up the convergence to the query concept. This scheme has proven to be effective in image retrieval by previous research work (Pluim et al., Burt and Adelson). In CBMIR, we develop three active learning methods based on different principles, which intentionally select images in each round of relevance feedback, aiming to maximally improve the ranking result (Catte et al. 1992, Dutt and Greenleaf 1996).

As pointed in previous section, $f^+ = A^+ y^+$ and $f^- = A^- y^-$ are the ranking scores obtained from positive and negative samples respectively. The final ranking score is $f^+ = A^+ y^+$. For an unlabeled image, $x_i f_i$ is in proportion to the conditional probability that $x_i$ is a positive sample given present labelled images: the larger $f_i$ is, the bigger the probability.

The first method is to select the unlabeled images with the largest $f_i$, i.e., the most relevant images, which is widely used in previous research work (Brown 1992, Jinhua et al. 2010, Josien et al. 2003, Medina et al. 2012). The motivation behind this simple scheme is to ask the physician to validate the judgment of the current system on image relevance. Since the images presented to the user are always the ones with the largest probabilities of being relevant, many of them might be labelled as positive, which will help the system refine the query concept; while the negative feedback images will help to eliminate false positive images (Meyer et al. 1997, Wen 2008).

The second method is to select the unlabeled images with the smallest $|f_i|$. Since the value of $f_i^+$ indicates the relevance of an unlabeled image determined by positive samples, while the absolute value of $f_i^-$ indicates the irrelevance of an unlabeled image determined by negative samples, a small value of $|f_i^+| = |f_i^+ + f_i^-|$ means that the image is judged to be relevant by the same degree as it is judged to be irrelevant, therefore, it can be considered an inconsistent one. From the perspective of information theory, such images are most informative (Medina et al. 2012).

The third method tries to take the advantage of the above two schemes by selecting the inconsistent images which are also quite similar to the query. To speak concretely, we define a criterion function:

$$c(x_i) = f_i^+ - |f_i^+ + f_i^-|$$

Unlabeled images with the largest value of $c(x_i)$ are selected for feedback. The criterion can be explained intuitively as follows: the selected images should not only provoke maximum disagreement among labeled samples (small $f_i^+ + f_i^-$), they must also relatively confidently be judged as relevant ones by the positive samples (large $f_i^+$). We justify this scheme as follows:

Generally speaking, since positive samples occupy a small region in the feature space and are surrounded by negative samples, to identify the true boundary separating the two classes of images with a small number of labeled samples, it is more reasonable to explore in the inconsistent region. If an image is far from all the labeled samples, it will have a small value for both $f_i^+$ and $|f_i^-|$ and a small $f_i^+ + f_i^-|$ accordingly, thus it is likely to be selected by the second scheme although it makes a small contribution to the refinement of the boundary. However, it is not likely to be selected by the third scheme according to equation. Therefore, this scheme is expected to outperform the second one.

All our experiments were run in Matlab version 7 environment. The evaluation is done using metrics such as sensitivity and specificity. The sensitivity measures the proportion of actual positives which are correctly identified, whereas, specificity measures the proportion of negatives which are correctly identified (Zhang et al. 2007, Zhang et al. 2006).

$$\text{Sensitivity} = \frac{\text{True Positives}}{\text{True Positives} + \text{False Negatives}}$$

$$\text{Specificity} = \frac{\text{True Negatives}}{\text{True Negatives} + \text{False Positives}}$$

Our retrieval system is trained using two different disease liver images, namely normal image and cyst image. During the retrieval phase, we run the query images through a series of trained classifiers to acquire the category to which the image belongs to. To demonstrate the performance of our system, for each diseased image category, we performed 3 active learning methods at testing phase and training through soft label SVM. The sensitivity and specificity shows positive samples and negative samples to each category of images (Meyer et al. 1997). This is one of the main advantages of using this method to retrieve similar images compared to other traditional approaches for biomedical image retrieval.

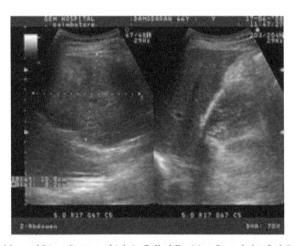

**Figure 45.** Normal Liver Image which is Called Positive Sample by Soft Label SVM.

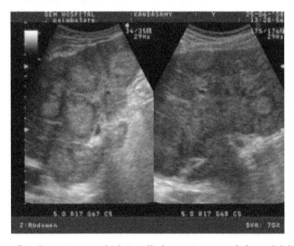

**Figure 46.** Cyst Liver image which is called negative sample by soft label SVM.

When the query image matches with the positive images shown in Figure 45, and belong to a specific class, it displays images from the pool belonging to the retrieved category with the highest similar value based on our ranking function calculation. And the negative cyst live image is shown in Figure 46. We also validated the classification performance by testing the system with noisy and distorted data and partially clipped images. Two categories of images if incorporated in the pool of test images may result in the specificity and sensitivity values. Compared to other traditional algorithms, the proposed system responds faster in providing retrieval features images for a given input image.

## 5.11 Hybrid Kohonen Self Organizing Map (HKSOM)

Hybrid Kohonen SOM uses a neighborhood function to preserve the topological properties of the input space. It has the ability to access the input patterns presented to the network, organize them to learn on their own the similarities among the sets of inputs, and cluster the input data into groups of similar patterns. Therefore, the result of this is a set of clusters, where each cluster belongs to a specific class. In this case, the system extracts information from unlabelled samples (Duda et al. 2001). A major problem in the unsupervised classifier in medical image classification is that of defining the similarity between two feature vectors and choosing an appropriate measure for it. Another issue is selection of the best matching unit algorithm that will cluster the vectors on the basis of similarity measure.

In the proposed HKSOM method, the network architecture consists of an input layer followed by a single competitive layer. Input layer consists of the six neurons where six feature values are extracted from the images. Each unit in the input layer is connected to single neuron in the competitive layer. Weights are attached to each of this connection, thus resulting in weight vectors, for each neuron, having the same dimensions as the input vectors. Initially, the weights are randomized. When a training vector X is presented, the weight vector of each neuron, i is compared with X. The neuron that lies closest to X, is the winning neuron. In this work the minimum of the Euclidean distance between the vectors X and Wj is taken as the winner neuron. The weight vector of the winner unit and its neighbours in the grid are adapted with the following learning rule. Here it operates in two modes: training and mapping.

- Training Phase: Map is built, network organizes using a competitive process, and it is trained by using large numbers of inputs
- Mapping phase: Newly generated vectors are quickly given a location on the converged map, easily classifying or categorizing the new data.

The systematic flow diagram for the proposed system—Hybrid Kohonen SOM with MLP is shown in Figure 47.

The procedure for the proposed approach is as follows:

1. Image Pre-processing is done by MLPND method to remove speckle noise and unwanted information from the input ultrasound liver images.
2. The Pathology Bearing Region (PBRs) of size 23 x 23 pixels for each kind of pattern is cropped and then fed into the feature extraction module.
3. Feature extraction is carried out by Gray level co-occurrence matrices (GLCM) and Fractal geometry (Hurst Fractal index).
   a) GLCM features represent the frequency of all possible pairs of adjacent gray level values in the entire image. Selected five GLCM features used in this work are contrast, cluster prominence, cluster shade, angular second moment, and auto correlation that have already been discussed in Chapter 4.

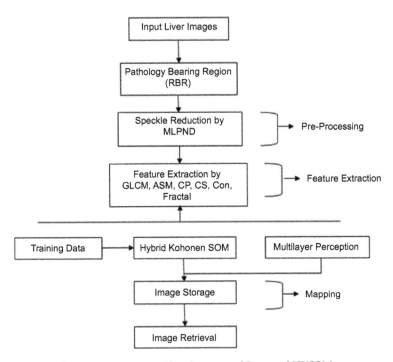

**Figure 47.** Systematic Flow Diagram of Proposed HKSOM.

b) Fractals provide a measure of the complexity of the gray level structure in a certain pathology bearing region, having the characteristics of self-similarity at different scales. This feature plays an important role in ultrasound image modality. Every texture, characterized through the intensity I, can be represented as a reproduction of the copies of N basic elements: One of the ways to express the fractal dimension is the Hurst Coefficient.

4. These features are finally passed to the hybrid Kohonen SOM classification module to determine the category of liver diseases. The hybrid Kohonen SOM is constructed by the following steps:
   a) Initialize each node's weight.
   b) Pick a random value from the training data and feed it to the HKSOM.
   c) Every node is examined to find the Modified Best Matching Unit (MBMU).
   d) The radius of the neighborhood around the BMU is computed. The size of the neighborhood value is decreases with each iteration.
   e) Each node in the BMU's neighborhood has its weights adjusted to become similar to the BMU. Nodes closest to the BMU are changed more than the nodes furthest away in the neighborhood.
   f) Repeat from step (b)–(e) for enough iteration for convergence.

5. The output from Kohonen SOM layer is passed to the Multi-Layer Perceptron (MLP) to classify the abnormal liver diseases by means of Kohonen map.

   i) Modified Best Matching Unit (MBMU) Procedure

      1. The Modified Best Matching Unit is calculated according to the Euclidean distance among the node's weights ($W_1$, $W_2$... $W_n$) and the input vector's values ($V_1$, $V_2$... $V_n$). This gives a good measurement of how similar the two sets of data are to each other.

$$\text{Dist} = \sqrt{\sum_{i=0}^{i=n} (V_i - W_i)^2}$$

      2. Size of the neighborhood is measured by using an exponential decay function (William et al. 2008). An exponential decay function that shrinks on each iteration until eventually the neighborhood is just the BMU itself.

$$\sigma(t) = \sigma_0 \exp\left(-\frac{t}{\lambda}\right)$$

3. The neighborhood parameter is defined by a Gaussian curve so that nodes that are closer are influenced more than farther nodes.

$$\theta(t) = \exp\left(-\frac{dist^2}{2\sigma^2(t)}\right)$$

4. The modified BMU is computed by modifying nodes weight which is already assigned for all neurons and its neighborhood function. Hybrid Kohonen SOM uses a neighborhood function to preserve the topological properties of the input space. The new weight for a node is the sum of the old weight, plus a fraction (L) of the difference between the old weight and the input vector is adjusted (theta) and divided by two based on distance from the BMU.

$$W(t+1) = W(t) + q(t)\, L(t)\, (V(t) - W(t))/2$$

5. The learning rate L is also an exponential decay function. This ensures that the SOM will converge.

$$L(t) = L_0 \exp\left(\frac{t}{\lambda}\right)$$

The $\lambda$ represents a time constant, and t is the time step.

## 5.12  Results and Discussion

This work has been carried out with 300 ultrasound liver images—60 cyst, 45 hemangioma, 66 normal liver and 41 hepatoma and 48 Cirrhosis and 40 fatty liver images were considered for the training phase. The size of the images was $256 \times 256$ pixels and the images were saved at 12 bits per pixel gray level. Initially, the speckles were removed from the ultrasound images by MLPND method which is shown in Figure 48; the training set for the classification algorithm was created by manually segmenting sample regions of six patterns: liver cyst, hepatoma, fatty liver, hemangioma, cirrhosis and normal liver. Among the 300 images, 210 PBRs of size $23 \times 23$ pixels for each kind of patterns were sampled for mapping as shown in Figure 49.

The Kohonen Hybrid SOM consists of an input layer, followed by a single layer of competitive neurons. In the case presented above the input layer consists of six neurons to which six feature values—contrast, autocorrelation, cluster shade, cluster prominence, angular second moment and fractal dimension are extracted from the despeckled image and given as

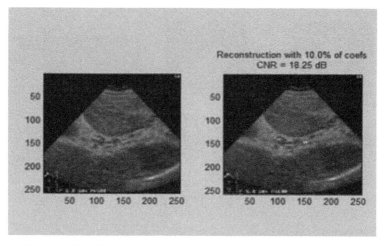

**Figure 48.** Speckle Reduction from Liver Hepatoma by MLPND Method.

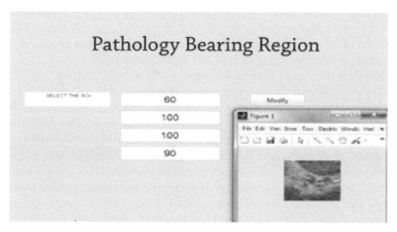

**Figure 49.** Sampling PBR of size 23 x 23 Pixels for Each Kind Liver Pattern.

input layer and followed by only one neuron at the competitive layer called the winning neuron. Weights are added to each neuron and calculate its best matching unit. The feature fractal plays an essential role in distinguishing between liver hepatoma and hemangioma which is shown in Figure 50.

Separate runs are given for each neighborhood and orientation and also for all three types of liver diseases. The output from the Kohonen layer is passed to the Multilayer Perceptron (MLP) to classify the normal and abnormal liver diseases by means of the Kohonen map. The training and

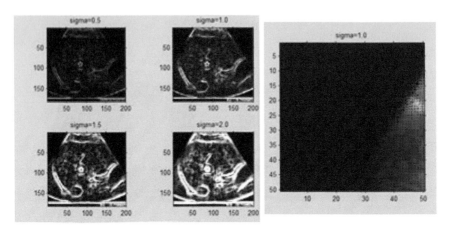

**Figure 50.** Feature Extractions from Liver Hepatoma by Fractal Method.

mapping phase are very unimportant in Hybrid SOM. Hybrid Kohonen SOM is fast and efficient with less classification error for the large dataset.

SOM is an unsupervised algorithm. It does not have any target values. It finds its neighbours through random weight assignment by calculating the best modified matching unit in the input value pairs.

The performance of Hybrid kohonen SOM is evaluated by Sensitivity, Specificity and Retrieval Accuracy for the test set. Area under the curve (AUC) is another performance metrics for classification and retrieval of medical images using machine learning techniques. The AUC using normalized units is equal to the probability that a classifier will rank a randomly chosen positive instance higher than a randomly chosen negative one. A reliable and valid AUC estimate can be interpreted as the probability that the classifier will assign a higher score to a randomly chosen positive example than to a randomly chosen negative example. Table 21 shows the performance evaluation of HKSOM by Sensitivity, Specificity, and Retrieval Accuracy rate.

From Table 21, it is inferred that the HKSOM achieved 96% retrieval accuracy rate. The comparison of the classification accuracy rate for the proposed methods HSVM-RF, multiclass SVM and HKSOM is discussed in the Table 22.

It is inferred from the Table 22, that the analysis of the obtained results for HSVM binary classifier with texture features is obtained, its accuracy rate is about 72.1%. But the accuracy rate of SVM multi class with minimal GLCM feature is obtained about 81.7%. Even though by adopting SVM multi class with five GLCM features, the retrieval accuracy rate is less

**Table 21.** Performance Evaluation of HKSOM with Sensitivity, Specificity and Retrieval Accuracy Rate.

| Liver Images | Count | Sensitivity | Specificity | Retrieval Accuracy Rate |
|---|---|---|---|---|
| Cyst | 60 | 95 | 96 | 95 |
| Hemangioma | 45 | 95 | 97 | 96 |
| Normal | 66 | 96 | 98 | 97 |
| Hepatoma | 41 | 97 | 92 | 95 |
| Cirrhosis | 48 | 95 | 100 | 97 |
| Fatty | 40 | 100 | 87 | 93 |

**Table 22.** Comparison of Proposed Methods for Liver Image Classification & Retrieval.

| Machine Learning Techniques | Feature Extraction Methods | Retrieval Accuracy Rate |
|---|---|---|
| SVM binary class with RF (HSVM-RF) | GLCM, First Order statistics Fractal geometry | 72.1% |
| SVM multiclass | Contrast, Auto correlation, Cluster prominence, Cluster Shade, Angular second momentum | 81.7% |
| Hybrid Self Organizing Map (HKSOM) | Statistical approach (Haralick's five features) and Model based approach (Fractal average dimension) | 96% |

and this leads to misclassification among focal and diffuse liver. But by combining five Haralick's texture features and fractal dimension (i.e., model based approach), the Hybrid Kohonen Self Organizing Map (HKSOM) classification of ultrasound liver images is achieved 96% correctly as shown in Table 22. However this result can vary with a large number of dataset containing different types of speckle noise. The key findings from this work are that classification of liver diseases like hepatoma and hemangioma are clearly distinguished with the help of SOM, but some samples of the cyst and fatty liver is grouped into one cluster which denotes the fuzzy nature of texture characteristics in liver dataset.

## 5.13 Summary

A Hybrid Kohonen SOM with MLP for the classification and retrieval of ultrasound liver images was implemented for a large dataset (300 liver samples). Six kinds of liver images were identified including normal, cyst, hepatoma, hemangioma, cirrhosis and fatty liver. Based on the experimental results, it is concluded that the fractal dimension play an important role in discriminating liver hepatoma and hemangioma. Hybrid Kohonen SOM enhances MBMU to preserve the topological properties of the input space. By using this machine learning technique, the classification error is reduced and the performance is increased by increasing the number of samples. The results obtained with Kohonen Hybrid SOM was evaluated with parameters namely Sensitivity, Area under Curve, and Specificity. The merits of Hybrid Kohonen SOM are as follows: (i) Accessing input patterns and clustering them into groups of similar patterns by modified Best Matching Unit (BMU). (ii) Handling higher dimensional medical data more efficiently. The limitation of our proposed system is as follows, it works well for the large number of dataset. So the proposed system can be extended for diagnosis of other types of liver diseases and also for the large dataset. The key findings from this work attains that classification of liver diseases like hepatoma and hemangioma are clearly distinguished with the help of HKSOM, but some samples of the liver diseases like cyst and fatty are grouped into one cluster which denotes fuzzy nature of texture characteristics in liver dataset.

## 5.14  Deep Learning Algorithm for Classification

The ultimate aim in deep learning algorithms is automating the withdrawal of representations like morphology or texture features from the data. Geert Litjens et al. (Geert Litjens et al. 2017) surveyed Deep learning algorithms and used a huge amount of unsupervised data to automatically extract complex representation. The important goal of deep learning is similar to the human being brain's ability to observe, analyze, learn and make decisions, especially for extremely multifaceted problems. Models based on shallow learning architectures such as decision trees, support vector machines and case based reasoning may fall short when attempting to extract useful information from complex structures and relationships in the input corpus.

A key fundamental idea of the Deep learning methods is distributed depictions of the data, in which a huge number of possible configurations of the abstract features of the input data are feasible, allowing for a compact representation of each sample and leading to a richer classifier. Deep learning algorithms are actually deep architectures of consecutive layers. Each layer feeds a nonlinear transformation on its input and gives a representation in its output layer. The purpose is to study a confused and abstract representation of the data in a hierarchical manner by passing the data through multiple transformation layers. The sensory data (texture features) is fed to the first layers. Consequently the output of each layer is provided as input to its next layer. It is essential to note that the changes in the layers of deep architecture are non-linear transformations which try to take out underlying explanatory factors in the data.

Deep learning architectures have the potential to generalize in non-local and total ways, supplying the learning models of patterns and relationships beyond immediate neighbours in the data which is represented in Figure 51. Deep learning means training network with many layers. Multiple layers work to build an improved feature space of medical image classification. The first layer learns 1st order of features like cluster prominence, auto correlation and homogenity. Second layer learns higher order features like LBP and SIFT (combination of first layer features and combination of edges, etc.). Some models are trained in an unsupervised mode and notice general features of the input space—helping multiple tasks related to the unsupervised instance. Final layer of transformed features are fed into supervised layers. And complete network is repeatedly tuned using supervised training of the entire set, and using the initial weightings adopted in the unsupervised phase.

To deal with translation invariance, each node in a feature map has the same threshold weights and every node joins to a different overlapping receptive pasture of the earlier layer. Thus each feature map completely searches the entire earlier layer. The output will be elevated at every node in the map corresponding to a receptive pasture where the feature occurs. Later layers could alarm themselves with higher order combinations of features and rough relative positions. Each calculation of a node's measure net value. Feature map in deep learning model is described convolution and it is based on the similarity to standard convolutions. Each node (convolution) is calculated for each receptive pasture in the earlier layer. During the training phase the corresponding weights are always tied to be equal. Thus there is a relatively small number of unique weight parameters to learn, although they are replicated many times in the feature map. Each feature map should be trained for a different translation invariant features from the SIFT feature. Since, after the first layer there are always multiple

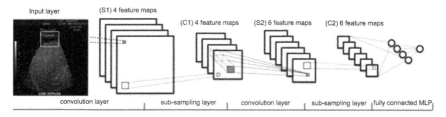

**Figure 51.** Deep Learning Algorithm with Multiple Layers.

feature (GLCM, LBP, SIFT) maps to connect to the subsequent layer, it is a pre-made individual decision as to which earlier maps the current convolution map obtains inputs from, and could join to all or a subset. Convolution layer causes the total number of texture features to increase.

## 5.15  Fuzzy Classifier (FC)

As the ultrasound liver dataset, namely liver cyst and fatty are fuzzy in nature, the above machine learning method—Support Vector Machine with Relevance feedback and Hybrid Kohonen Self Organizing Map cannot provide accurate results. Hence this part of the work is modeled by using Fuzzy classifier. Fuzzy classifier plays an important role in dealing with uncertainty when making decisions in medical applications (Kuncheva Ludmila and Fried Rich Steimann 1999). Classification of medical image objects is based on association of a given object with one of the several classes (i.e., diagnoses). Each object does no longer belong entirely to only one class but rather to all classes with different membership values. Fuzzy classifier provides a good framework for medical image analysis. The main advantage of this framework is to represent and manipulate the fuzzy data contained in medical images and thus it provides flexibility in presenting extracted knowledge to clinicians and radiologists.

- System Overview for FC

This section deals with images of ultrasound liver that contain presence of cyst or fatty deposits. All these images are gray scale images, so it is found that texture would serve as a better feature to identify the presence of disease. The size of cyst varies from person to person there is no fixed range to identify the disease. Hence it is essential to identify the presence of cyst or fatty deposit accurately. To overcome the problem mentioned above, fuzzy classifier is employed to eliminate the imprecision involved in classifying liver cyst and fatty. The Systematic flow diagram for fuzzy classifier is as shown in the Figure 52. It consists of three main steps: (a) Extraction of Texture (b) Fuzzification by Fuzzy Membership Function and (c) The classification of the liver images by comparison of membership function and retrieval using a fuzzy classifier.

Procedure for US liver image classification and retrieval by fuzzy classifier:

1. Noise free ultrasound liver image is fed in to the fuzzy classifier system.
2. Calculate feature extraction by using entropy (GLCM feature) for texture analysis.

166

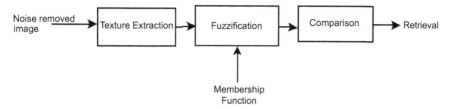

**Figure 52.** Systematic flow Diagram for Fuzzy Classifier.

3. Fuzzify the values by triangular membership function.
4. Apply fuzzy classifier to compare the query value with the database and retrieve its result.
5. Three sets of ultrasound liver images were used in this experiment: Normal, Cyst and Fatty. All ultrasound images are captured with 512 x 512 pixels, 256 gray-level resolutions, with image gain of 18–20 DB, image depth of 18–21 cm and mechanical index is between 0.8–1.2. Texture analysis algorithms were applied in each image on a 32 X 32 pixels pathology bearing region (PBR) selected in a systematic way so as to avoid deviation in image statistics. They have the following principles:

   a. All liver images are taken by the same radiologist, using the same ultrasound system setup.
   b. Ultrasound system settings that affect image texture are kept the same.
   c. PBR is selected by expert radiologist so as to contain only liver parenchyma (normal or abnormal) with no major blood vessels. PBR is selected along the focussing area of texture present in each image.

In the following section, the techniques used for feature extraction and classification are presented.

   i. Liver Image Texture Analysis

Three different image texture analysis techniques were used Entropy, GLCM, and Fractal. Sometimes images with fatty deposits might be unpredicted as images with cysts because of the similar appearance in case of manual diagnosis. Therefore there is a need for exact texture feature in order to classify the images and to facilitate easy retrieval. The texture of an image can be described using:

- GLCM
- Entropy
- Fractal

## 5.15 Entropy

Of these, entropy is taken as the measure to describe the texture of an image. Entropy becomes the self-information of a random variable. In general, entropy is a statistical measure of randomness that can be used to characterize the texture of the input image.

### 5.15.2 Fractal Dimension Texture Analysis (FDTA)

The last technique used to quantify texture in ultrasound images is based on the Fractional Brownian Motion (FBM) theory as presented by Mandelbrot (1982). FBM is a non-stationary stochastic process which can be described by a single parameter, its fractal dimension D, where D is equal to E+1-H. The parameter H is the so-called Hurst coefficient and E+1 is the Euclidean dimension of the embedding space of the fractal. FBM and the corresponding D and H parameters can be used to describe the roughness of different surfaces. Similarly the work uses Hurst coefficient for levels 32 X 32 (original region) H5 and 16 X 16 H4 to characterize ultrasonic liver images. The fractal dimension texture analysis is based on the fact that a rough surface gives a small value of H whereas a smooth surface gives a large value of H.

### 5.15.3 Fuzzification by Fuzzy Membership Function (FMF)

The difference between crisp (i.e., classical) and fuzzy sets is established by introducing a membership function. A membership function is a mathematical function which determines the degree of an element membership in a fuzzy set. It may be defined as a characteristic function of a fuzzy set that assigns to each element in a universal set a value between 0 and 1.

A membership function (MF) is a curve that defines how each point in the input space is mapped to a membership value (or degree of membership) between 0 and 1. The input space is sometimes referred to as the universe of discourse. (Jin Zhao and Bimal 2002) defined that the membership function is similar to the indicator function in classical sets. For any set $X$, a membership function on $X$ is any function from $X$ to the real unit interval

[0, 1]. There are many types of Membership Functions of which three of them are explained in detail below:

a) Triangular Membership Function
b) Trapezoidal Membership Function
c) Gaussian Membership Function

*a) Triangular Membership Function (TMF)*

A triangular membership function is described by three parameters a, b and c. The expression is given below:

$$f(x,a,b,c) = \max \left\{ \min \left( \frac{x-a}{b-a}, \frac{c-x}{c-b} \right), 0 \right\} \tag{57}$$

where a and c represent the feet of the triangle and b represents the peak of the triangle. Triangular membership function is shown in Figure 53.

*b) Trapezoidal Membership Function*

A trapezoidal membership function is described by 4 parameters a, b, c and d. It is given by the expression:

$$f(x,a,b,c,d) = \max \left\{ \min \left( \frac{x-a}{b-a}, \frac{d-x}{d-c} \right), 0 \right\} \tag{58}$$

Here a and d represent the feet of the trapezoid and b and c represent the shoulders. Trapezoidal membership function is shown in Figure 54.

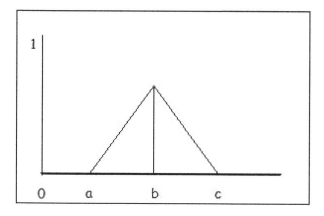

**Figure 53.** Triangular Membership Function.

*c) Gaussian Membership Function*

Gaussian Membership Function can be given by the formula:

$$f(x,a,b,c) = e^{\frac{-(x-c)^2}{2\sigma^2}} \tag{59}$$

Here c represents the distance from the origin and σ indicates the width of the curve. Gaussian membership function is shown in Figure 55.

Among the above mentioned Membership Function, the Triangular membership function is used in this book for simplicity of computation which thereby reduces the time of execution. In a triangular membership function, the peak of the triangle is given by averaging the lower and upper limits. Therefore, the values are fuzzified and values ranging between 0 and 1 are obtained. The entropy value of the query image is fuzzified at ranges of normal and fatty image. The maximum of these values reveals the class of the query image.

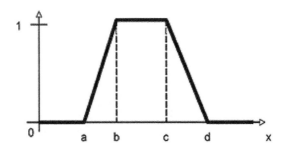

**Figure 54.** Trapezoidal Membership Function.

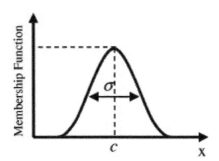

**Figure 55.** Gaussian Membership Function.

170

### 5.15.4 Fuzzy Classifier

In order to differentiate the fuzzy nature of Cyst and fatty liver, the work uses a fuzzy classifier. Out of above three texture analysis methods, a range of entropy values, mean and STD for each category of images are considered for classification of liver diseases that is shown in Table 23. After constructing the texture values from entropy, decision probabilities are computed based on the density of membership values for each region and the respective performance in the selection of the pathology bearing region. Through discretization of the membership function, a probabilistic function is created that establishes a correspondence between membership values in a specific region and the probability of correct classification. The construction of the probability models is based on the search for ranges of membership values that have more chance to lead to the successful selection of a region. More specifically, considering class region I, the interval [0, 1] of membership values is divided into a number L, of equal-size cells. To each cell $(v-1, ...L)$ assign a probability value $P_i^v$ computed as the percentage of the training patterns belonging to region I that have their membership value in the cell v. For each image in the liver dataset, the entropy values are found and fuzzification is performed using the Triangular Membership Function. The difference between the fuzzified entropy value of the submitted query image and the fuzzified entropy value of each image is found. The matching range is found (the category in which the query image falls) and the corresponding script is executed to find the nearest similar image. Minimal difference indicates more similarity. Therefore, the image with the minimal difference is retrieved.

**Table 23.** Entropy, Mean & STD Ranges of Liver Images.

| Type of Image | Entropy Range | Mean | STD |
|---------------|---------------|--------|--------|
| Normal liver | 4.3–5.8 | 0.4103 | 0.1030 |
| Liver cyst | 4.1–6.2 | 0.6122 | 0.2122 |
| Fatty liver | 3.3–5.4 | 0.6890 | 0.2872 |

## 5.16 Results and Discussion

To evaluate the performance of the proposed fuzzy classifier technique, the research uses a total of 150 images (Normal, Cyst and Fatty). As described previously, 32 X 32 pixels PBR were selected in each image and the corresponding texture features were calculated. These features have been shown to contain useful information for characterizing liver tissue pathology. From this, entropy is selected for this research work. The entropy range of normal, cyst and fatty liver were found to be ranging from 4.3–5.8, 4.1–6.2 and 3.3–5.4 respectively. From the entropy values it is concluded that for an input image that has the entropy value of 4.1, it cannot be decided whether it has a cyst or fatty deposits. Therefore, fuzzification on the entropy value of the image is done by using Triangular Membership function. In order to handle with the imprecision and overlapping bounds, fuzzy classifier is implemented, that reduces the misclassification error. But the retrieval time is increased by using this system as is shown in Table 24. The dataset of ultrasound liver images is shown in Figure 56. For the input query cyst, the corresponding cyst images are retrieved by using fuzzy classifier ad is shown in the Figure 57. For fatty liver, the retrieval of fatty liver image is shown in the Figure 58.

**Table 24.** Performance of Retrieval Time for Triangular Membership Function.

| Query Image | Retrieval Time (Sec) |
|:-----------:|:--------------------:|
| Cyst | 3.63 |
| Fatty | 2.21 |
| Normal | 2.37 |

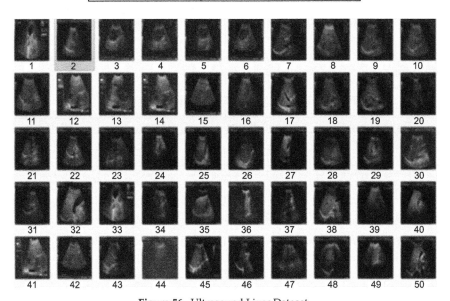

**Figure 56.** Ultrasound Liver Dataset.

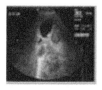

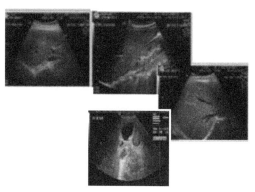

**Figure 57.** Retrieval for the Cyst Query Image.

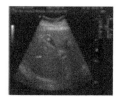

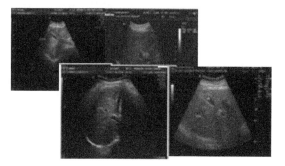

**Figure 58.** Retrieval for the Fatty Query Image.

## 5.17 Summary

Computer assisted characterization of ultrasonic liver images using image texture analysis techniques and fuzzy classifier has been tested and evaluated. The result shows that the fuzzy nature of cyst and fatty liver is classified more accurately with the help of fuzzy classifier. In particular, this system has adopted triangular membership function for fuzzification. Triangular membership function is used in this work for simplicity of computation which thereby reduces the time of execution. Experimental results indicate that the proposed method is effective in terms of the rate of correct classification and has the ability of overcoming the difficulties arising from the problem of overlapping classes. Furthermore, the results of this work have demonstrated the ability of using image texture analysis (entropy, mean and STD) technique to extract features that characterize liver tissue abnormalities. It is possible to conclude that Entropy when used as a feature in Fuzzy classifier gives better results for classification.

## 5.18 Conclusion

To conclude, most of the machine learning algorithms for today's medical imaging system have originally been both spawned and incubated to provide well designed and efficient solutions for classification of pathological lesions and cysts. This book proposes new methods for speckle reduction, image registration, feature extraction, classification and retrieval algorithms facilitated automated diagnosis, so as to improve the diagnostic accuracy of the ultrasound of liver diseases.

Three different approaches namely Support Vector machine with binary class and multiclass, Hybrid Kohonen SOM and fuzzy classifier have been investigated individually. SVM is a supervised learning technique, used for ultrasound liver image classification and retrieval. For supervised learning, a set of training data and category labels are available and the binary classifier is designed to classify only positive and negative samples. Multiclass SVM has been suggested as a solution for classification of liver images when the number of classes is quite high. Relevance feedback algorithms incorporated with SVM helps to improve the accuracy of retrieval of relevant images from a database. SVM with RF is used in retrieval system for two reasons: (1) to reduce the semantic gap (there can be a big gap between high level concepts perceived by the user and low level features that are used in the system) and (2) to subject similarity of human perception.

For a large liver dataset with different stages of liver diseases, the Hybrid Kohonen Self Organizing Map (SOM) algorithm utilizes the concepts of competitive learning. Competitive learning is an adaptive process in which the neurons gradually become sensitive to different input categories. HKSOM is different in the sense that they use a neighborhood function MBMU, which is used to preserve the topological properties of the input space. Diseases like hepatoma and hemangioma are classified effectively. Sometimes images with fatty deposits might be unpredictable from images with cysts because of their similar appearance in texture characteristics and their topological properties are grouped into one cluster. Hence to overcome these kinds of mispredictions it is essential to develop a system that would classify the images into proper categories. Fuzzy classifier provides a good framework for medical image analysis. The main advantages of this framework is that it provides a way to represent and manipulate the fuzzy data contained in medical images, and thus it provides flexibility in presenting extracted knowledge to clinicians and radiologists. Fuzzy classifier uses Triangular Membership Function (TMF) to classify diseases like cyst and fatty liver which are fuzzy in nature.

The result has been evaluated at each phase by the doctors in the relevant field. In image classification and retrieval, the Relevance feedback algorithm is incorporated with SVM to improve the accuracy of retrieval of relevant images from the database and reduce the semantic gap. Besides, image registration in HSVM-RF is used to monitor the growth of the liver disease and reduce the redundancy of retrieval images from the database. Hybrid Kohonen SOM uses Modified Best Matching Unit (BMU) to preserve the topological properties of the liver diseases like cirrhosis, hepatoma and hemangioma. Fuzzy classifier used triangular membership function to classify diseases like cyst and fatty livers which are fuzzy in nature. The future scope of the classification and retrieval phase could be extended for further classifying the diseased liver and labeling it with more number of diseases. A complete web interface system integrating the challenges in handling medical image modalities could be presented. Fuzzy based system proposed could be evaluated with other membership functions and their performance could be compared for analysis.

The existing challenges under Classification and Retrieval of ultrasound liver images include speckle noise, semantic gap, computational time, dimensionality reduction and accuracy of retrieval images from large dataset. All these issues are critical in nature and they have been addressed in each phase of the work. In this book, an attempt is made to address these issues and appropriate methods are proposed at each step that have been validated with the medical specialist in the relevant field.

Speckle noise reduction is completed by concentrating on improving edge information and retains subtle features by Modified Laplacian Pyramid Nonlinear Diffusion (MLPND) in the first step. The existing methods can be applied only to log compressed scan data so they fail to retain small features like cyst and lesions present in ultrasound liver images. The proposed MLPND preserves edges and small structure while maximally removing speckle and also can be applied to raw ultrasound images. Thus, it has the potential to improve the diagnostic capability of current ultrasound imaging and to enhance the performance of some high level image processing tasks in the ultrasound imaging systems. The performance of this approach has been tested with a 300 dataset. The result had an accuracy of 95%. The CNR value is able to provide the conclusion that the image clarity is not affected.

As the second step, an efficient method for monitoring the growth of liver diseases by mutual information based image registration with S-Mean filter is suggested. Two optimization techniques—DIRECT and Nelder Mead are applied. The computation time of mono modal liver image registration is reduced by DIRECT method compared to Nelder-Mead method. The result demonstrates that the mutual information based image registration allows fast, accurate, robust and completely automatic registration of mono modality medical images and is very well suited for clinical applications. The strength of mutual information value reduces the redundancy of mis-registered images during classification.

As the third step, Haralick's textural features and Fractal features are used to extract salient gray-level features from the Pathology Bearing Region (PBR) of liver images. The performance parameters are evaluated and the results proved that diagnosis can be completed with minimal features which help in dimensionality reduction. Five of the twelve features (contrast, cluster prominence, cluster shade, Angular second momentum, and Auto correlation) used are found by Correlation Feature Selection to be the strongest describers of a texture which can be observed from feature extraction results. Experimental results shows that the analysis of the obtained results suggested that diseases like Cyst, Fatty Liver, and Cirrhosis can be diagnosed with only five features out of twelve features belonging to Haralick's textural features. Texture properties coupled with fractal produced good accuracy rate and help to diagnosis liver hepatoma from liver hemangioma.

As the final step, three machine learning algorithm are investigated for classification and retrieval of ultrasound liver images. Relevance feedback algorithm is incorporated with SVM to improve the accuracy of retrieval of relevant images from the database and the reduce semantic gap. The analysis of the obtained results for HSVM-RF with GLCM, first order statistics and fractal are obtained; its accuracy rate is about 72.1%. Besides, RF in liver classification, image registration plays an important role in retrieval. It reduces the redundancy of retrieval images. However, the accuracy rate of SVM multi class with selected five GLCM features obtained is about 81.7%. Even though by adopting SVM multi class with selected five GLCM features, the accuracy rate is less and this leads to misclassification among focal and diffuse liver. This work is extended to the unsupervised learning algorithm.

An unsupervised machine learning algorithm—Hybrid Kohonen SOM uses Modified Best Matching Unit (MBMU) to preserve the topological properties and Multilayer perceptron is used to classify liver diseases like cirrhosis, hepatoma, hemangioma, cyst and fatty. By using this machine learning technique, the classification error is reduced and the performance is increased by increasing the number of samples. The results obtained with HKSOM is evaluated with parameters namely Sensitivity, Area under Curve, and Specificity. The accuracy rate of Hybrid kohonen SOM with GLCM and fractal texture feature is about 96%. However this result can vary with a large number of dataset containing different types of speckle noise. The merits of Hybrid Kohonen SOM lies in accessing input patterns and clustering them into groups of similar patterns by MBMU, but some samples of the liver cyst and fatty is grouped into one cluster which denotes the fuzzy nature of texture characteristics in the liver dataset.

Hence to overcome these kinds of mis-predictions it is essential to develop a system that would classify the images into proper categories. Fuzzy classifier provides a good framework for medical image analysis. Fuzzy classifier used triangular membership function to classify diseases like cyst and fatty which are fuzzy in nature. The result has been evaluated at each phase by Doctors in the relevant field. This work in overall proposes a more automated decision making system. More focus in this research has been in classification approaches for automated decision making system. Retrieval is part of the classification mechanism in each phase of the research, without concentrating on content based retrieval algorithms.

To conclude, most of the machine learning algorithms for today's medical imaging system have originally been both spawned and incubated to provide well designed and efficient solutions for classification of pathological lesions and cyst. This book proposes new methods for speckle reduction, image registration, and feature extraction; classification and retrieval algorithms facilitate automated diagnosis, so as to improve the diagnostic accuracy of ultrasound liver diseases.

## 5.19  Security in Medical Images

Medical Images are used to diagnose and detect the abnormalities in the human body. In this chapter, we have focused on ECG (Electrocardiography) signals that play a vital role in diagnosing Cardio Vascular Disease. CVD (Cardio Vascular Disease) is most prevalent among because of changing lifestyle, age, Tobacco, etc. According to the WHO (World Health Organization) report around 17 million people die each year because of CVD, which represents 31 percentage of the overall death rate. Most of the ECG signals along with the patient's sensitive data are transmitted through the internet for the e-healthcare system; it should be made sure that those details are secure during the transmission. Here the ECG Steganography technique has been used in which the patient details and diagnosis report will be converted into QR (Quick Response) code. Since QR code has been used to store the patient data and diagnosis details, QR code of 177 pixels can be used to store few pages of data. And it will be useful for the e-healthcare system to easily read the digital image using a sensor. Dual Tree Complex Wavelet Transform (DTCWT) used for the decomposition of the ECG signal and QR code will be embedded into it. The novelty of our study is the use of QR code, since the size of the watermark remains as the major drawback. When large amount of ECG signals are selected, processing time will increase which can be reduced by the use of Big data.

In the medical domain, there has been a huge advancement in the recent years due to the tremendous increase in the use of internet. Thus it falls to the e-health care system to send patients information such as patient name, identification number, age, location, previous medical report, diagnosis report, etc. through the internet, leading to a new era of digital images that are sent to various hospitals for further diagnosis. While sending this information, we should consider the fact that the information can be modified, tampered, lost, falsified, etc. during transmission. So, care should be taken to prevent these kinds of attacks during transmission.

### 5.19.1  *Importance of Electrocardiogram (ECG) Signal*

Disorders such as Myocardial infarction and myocardial ischemia are the major reason for heart attacks and ECG signals play a vital role in identifying the symptoms of these disorders. ECG is the first line of investigation choice for any patient coming with chest pain. It determines the rhythm, rate, cardiac axis and electronic conduction through the heart by PQRST complex.

Normal ECG signal as shown in Figure 59, should have the PR interval in the range of 120–200 ms, QT interval up to 440 ms and QRS duration up to 120 ms.

Signals can be used for the identification of diseases such as myocardial infarction which refers to the death of cardiac tissue (i.e., blood flow stops for a particular time), myocardial ischemia which occurs due to the lack of oxygen and leads to the coronary artery obstruction whereas synus rthyum represents the normal signal. Any abnormalities in the heart will be easily and instantly picked up by ECG so that treatment for the patients can be taken without any delay. RR interval determines the heart rate. ST elevation indicates that there is myocardial infarction whereas ST depression indicates that there is myocardial ischemia. Q wave inversion indicates old myocardial infarction. Location of this changes in lead I, lead II and lead III, avF, avL and avR, V1, V2, V3, V4, V5 and V6 indicates the location of the pathology whether it is an anterior wall, inferior wall, etc. in the heart.

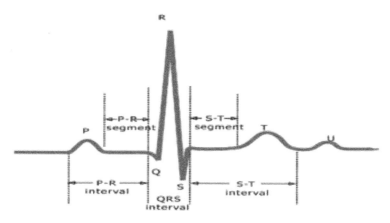

**Figure 59.** ECG Signal.

### 15.19.2 Providing Security in e-Health Care System

Security is very essential within e-Health care system since it will have a direct impact on the patient rights, quality of patient care and responsibilities of the healthcare professionals. Doctors can provide the correct diagnosis only when they are provided with the proper medical report and PHI (Protected Health Information) details (such as name, age, date, medical record number, email address, biometric identifier, Identification number, etc.) about the patients. Any change in these details may have an adverse effect in the patient's treatment.

There are various acts that have been adopted in the e-Health care system to ensure security in the medical domain. They are HIPAA (Health Insurance Portability and Accountability) act, IT (Information Technology) act, Personal Health Information Protection Act (PHIPA) which was established in 2004 and Personally Controlled Electronic Health Record (PCEHR) Act, 2012. According to the HIPAA act (Zhang et al. 2007), privacy regulations have to be maintained which deals with the Protected Health Information (PHI) as mentioned above and Confidentiality has to be taken care while dealing with the security of the data.

The Information Technology Act, 2000, has various updates based on the latest technological innovations. The main objective of IT act is to deal with cybercrime. In 2011, IT Act introduced the rule 'sensitive personal data' for the first time in India. PHIPA deals with the set of rules for the collection and prevention of the PHI details whereas PCEHR is a shared e-health care setup to provide security to the people's medical history details. In order to provide security, most widely used techniques are Cryptography and Steganography. Cryptography refers to secret coding. It is the art of transforming a text into some other form so that only the legitimate user will be able to understand the text. Another method is Steganography which deals with the embedding of secret messages such as text, audio, video, etc. into the cover signal.

Here a study about ECG Steganography has been carried out; it uses the watermarking methodology to embed the PHI details into signals. So that both PHI details and ECG will be transmitted securely. The significance of steganography lies in the fact that only the sender and receiver knows the fact that PHI is embedded in signals, others will be unaware of it and consider it as an ECG signal.

## 5.20 Basic Concepts in Watermarking Scheme

Watermarking is a sub-domain of steganography which deals with embedding the secret message into the cover signal, here both the cover signal and secret message has to be safe. As shown in Figure 2, there are various classifications in the watermark technique such as the methodology to be used, reversibility of the cover signal, watermark to be used and the human perception of the watermark. Based on the reversibility of the cover image there are two types. In Reversible watermarking, the cover image can be completely recovered from the watermarked image. Irreversible watermarking would mean that there will be a distortion of the cover image after the extraction of the data from the watermarked image. Text, Audio, Video and Image files can be used for the watermarking purpose.

The two different ways in which the watermarking can be applied are visible and invisible watermarking. Invisible watermarking is when the watermark will not be visible to human perception. In visible watermarking, after embedding the watermark into the cover image it will be visible to all. For example, a company logo will be embedded into all the documents that they are using. Based on the detection algorithm, two types of watermarking are there—blind and Non-blind. In blind watermarking, there is no need for the original image for the extraction process. For Non-blind watermarking, original image is needed at the receiver end for the watermark purpose. Robustness is a property which specifies the ability of the stego image to resist the attack. A robust watermark is that the stego image can withstand various attacks. Even for a slight modification, a fragile watermark will be easily destroyed. Semi-fragile will resist only a few attacks. There are two steps in watermarking.

*a) Watermark Embedding*

Watermark data must be embedded into the cover file using either the Spatial or Transform domain technique. This step is done at the sender side. After the watermark, has been embedded into the cover file it will be sent to the receiver. Now the cover image is referred to as Stego image.

*b) Watermark Extraction*

In this step, the watermark data must be extracted from the Stego image. The receiver should be aware of the technique used for watermarking so that he can use the respected Stego image and technique to extract the original image. Sometimes Stego image will be similar to the original image so that the user can use the same image as the input for further analysis.

## 5.21 Literature Survey

Jero (Abd-Elmoniem et al. 2002) proposed an approach to perform ECG steganography using curvelet transform. That is, using the transform domain technology to perform steganography. He used the MIT-BIH normal sinus rhythm database of ECG signals for the experiment purpose. Here the Fast Discrete Curvelet Transform has been used for the conversion of 1D signal to 2D signal and curvelet coefficients are calculated. Patient's data converted into binary form and threshold selection has been performed to select the coefficients (that are not representing the curve) and the n*n sequence method has been applied to the coefficients that are selected, to avoid overlapping. This method leads to the better extraction of the watermark bits. Inverse of the watermark embedding technique is used to extract the watermark bits. BER (Bit Error Rate) is observed to be zero for all the test cases which proves that embedded patient data can be extracted without any loss. PSNR (Percentage Signal to Noise Ratio) value of 75% has been observed for the small data size. And it is decreased at a rate of 10% for 1.5 increase of data size.

Chen (Amores and Radeva 2005) has implemented ECG Steganography in the MIT–BIH database. He performed a comparison between the transform domain based watermark techniques such as DWT (Discrete Wavelet Transform), Discrete Fourier Transform (DFT) and Discrete Cosine Transform (DCT). Blind watermarking has been simulated on the ECG signal using patient data such as name, age and ID, etc. as the watermark data. Using DFT, 7-level decomposition performed on the ECG signal using Haar filter. Watermark is embedded in the lowest frequency coefficients in all the techniques. This experiment has been implemented in MATLAB for about 4096 samples. Based on the performance measure (RMSE and rRMSE), DWT and DCT shows better performance when compared to DFT. Here they have not considered the size of the watermark to be used.

Ibaida (Aschkenasy et al. 2006) proposed a wavelet-based steganography technique with four steps. He used encryption and the scrambling technique in order to provide security to the patient data that has been sent through the network. Here five-level decomposition of the ECG signal has been done using the Discrete Wavelet Transform (DWT) technique and the patients' data are embedded into the LSB bits of the sub-band. In the watermark extraction, will be the reverse process of the embed process, in which decomposition of the watermark signal will be done initially and the scrambling matrix will be used to extract the bits in the proper order. Finally shared key will be used to decrypt the data. He has used the Percentage Residue Difference (PRD) and Wavelet Weighted PRD to measure the level of distortion, which is found to be less than 1%.

Jero (Brown et al. 1997) has proposed a robust ECG Steganography method in order to embed the Bose–Chaudhuri–Hocquengham (BCH) error—correcting coded patient data in the ECG signal using DWT–SVD technique. Here DWT used to decompose the signal and SVD is applied to the selected sub-band in order to embed the watermark. Experiment has been carried by watermarking the ECG signal with and without using the BCH code, and it is observed that the usage of BCH code corrects the bit error by 28.4% but the PSNR value has been reduced by 10.5%.

Wang (Burt and Adelson 1983) proposed two methods to perform ECG reversible data hiding. In the first method, PPE-HS technique has been used. PPE (Predicted Error Expansion) is an improvisation of Difference Expansion (DE) in which the message will be embedded into the prediction errors. PPE has been used to predict the sharp histogram and Histogram Shifting (HS) used to embed the secret data by modifying the histogram. In the second method, Unified ECG embedding–scrambling method has been used for reversible data hiding. The only difference in this method is to predict all the points and the data will be embedded into the respective points. It has been observed that the distortion is less than 1% between the watermarked and original ECG signal.

## 5.21 Proposed Methodology

Our proposed idea is to convert the patient data such as name, age and other PHI information along with the diagnosis report into QR (Quick Response) code. Since QR code has been observed as the best optimization technique to store data. And it was found that few pages of data can be embedded into 177 pixels of QR image. QR image is observed to be more error tolerant compared to other methods (Bui 2014). QR image will be converted into binary form. Based on the criteria n-level of decomposition takes place on the ECG signal using the DT-CWT (Dual Tree Complex Wavelet Transform) and SVD (Singular Value Decomposition) is applied on the coefficients which will provide us the singular values of the ECG signal. Similarly, SVD applied on the QR code, to get the singular values of the QR. Now the singular values of the ECG will be replaced with the singular values of the QR. And the watermarked signal will be transmitted.

At the receiver end, reverse process will be carried in order to extract the QR code which can be directly read with the help of sensors. Here we are going to perform ECG Steganography on the MIT-BIH database. And performance metrics such as PSNR, BER and KL will be used to measure the distortion level of the watermarked signal vs. the original signal. We have to analyze whether this technique will maintain the robustness and fidelity of the ECG signal.

1. Preprocessing
    i. ECG signal are collected from the international standard database (MIT-BIH database). MATLAB code to read the input signal that has been collected from MIT-BIH database.
    ii. Encryption of the PHI details. In order to improve the security of the PHI details, they will be encrypted before converting to QR code.

Encryption of the PHI details was done using AES 128 encryption algorithm.

   iii. Conversion of data to QR code. All the PHI information such as patient name, date of birth, age, medical report, etc. are collected from the hospital and converted to QR code.

2. Watermark embedding
   i. Decomposition using DT-CWT
      1. Here the QR code will be considered as image and DT-CWT 2D will be applied to the QR image.
      2. ECG signals from MIT-BIH database are read using the MATLAB code.
      3. DTCWT – 1D decomposition of the signal will be carried out.
      4. Multi-level Dual Tree complex wavelet transform are applied to the QR image and ECG signals.
3. Apply SVD: SVD will be applied to the HH sub band of the ECG signal, since the HH sub band has been observed as having least important information. Embedding details in this sub band will not lead to much distortion to the signal. Similarly, SVD will be applied to the LL sub band of the QR. Singular value of the ECG signal will be replaced with the singular value of the QR image.
4. Embed watermark
5. Inverse DT-CWT
   1. After the Singular Value Decomposition and the SV of the ECG signal are replaced with the SV of QR, inverse of the SVD will be performed by using the formula.
   2. And the resultant will be embedded into the level 2 HH sub band as mentioned earlier.
   3. Now inverse of the Dual Tree Complex Wavelet Transform will be performed to get the Stego ECG Signal.
6. Watermark extraction
   i. Input will be the Stego ECG signal and the DT-CWT will be applied to the ECG signal.
   ii. And the HH sub band will be subjected to SVD to obtain the QR code.

Here we have applied ECG Steganography in which the PHI details will be embedded into the ECG Signals. 1D ECG signal has been subjected to Dual Tree Complex Wavelet in order to decompose the signal and then the Singular value decomposition has been applied to get the singular vector of the ECG signal. Let us take it as S1. And the PHI (Personal Health

Information) details are gathered from the patients and they are stored in the document.

Encryption will be applied to the PHI details. This level of encryption will improve the security of the PHI details. Even when the attackers get to know the level of decomposition they can't easily access the PHI details and it will be converted to QR code. SVD will be applied to the QR code, and the embedding process SV of the Signal will be replaced with the SV of the ECG signal and then the inverse decomposition technique will be applied. At the receiver end, the same procedure will be carried out, after the decomposition the data will be extracted from the High-level sub bands and SVD will be applied to the coefficients. For the performance analysis, original ECG signal and the Stego ECG signal are given as the input and the performance metrics such as PSNR, MSE and BER are calculated.

Decomposition has been carried out using the DT-CWT technique. This approach has two major advantages. Firstly, it overcomes the widely occurring attacks in watermark methodology and improves the size of data that can be embedded inside the ECG signal by converting the data and other PHI into QR code. Main focus after performing steganography is that doctors should be able to provide diagnosis based on the Watermarked ECG signal, by using this technique less distortion has been observed in the watermarked ECG signal. Selected Transform based Watermark technique overcomes the watermark domain based issues (such as lack of directionality, shift variance, oscillations, etc.). The chosen Optimization methodology (QR code) rules out the distortion issue due to watermark size.

## 5.22  Advantages to Healthcare

By medical image processing, combining and effectively using big data, healthcare organizations ranging from single-physician offices and multi-provider groups to large hospital networks and accountable care organizations stand to realize significant benefits. Potential benefits include detecting diseases at earlier stages when they can be treated more easily and effectively; managing specific individual and population health and detecting health care fraud more quickly and efficiently. Numerous questions can be addressed with big data analytics.

Certain developments or outcomes may be predicted and/or estimated based on huge amounts of historical data, such as length of stay (LOS); patients who will choose elective surgery; patients who likely will not benefit from surgery; complications; patients at risk for medical complications; patients at risk for sepsis, MRSA or other hospital-acquired illness; illness/disease progression; patients at risk for advancement in disease states; causal factors of illness/disease progression; and possible co-morbid conditions (EMC Consulting). McKinsey estimates that big data analytics can enable more than $300 billion in savings per year in U.S. healthcare, two thirds of that through reductions of approximately 8% in national healthcare expenditures. Clinical operations and R & D are two of the largest areas for potential savings with $165 billion and $108 billion in waste respectively. McKinsey believes big data could help reduce waste and inefficiency in the following areas:

*a) Clinical Operations*

Comparative effectiveness research to determine more clinically relevant and cost-effective ways to diagnose and treat patients.

*b) Research and Development*

(1) Predictive modeling to lower attrition and produce a leaner, faster, more targeted R & D pipeline in drugs and devices; (2) Statistical tools and algorithms to improve clinical trial design and patient recruitment to better match treatments to individual patients, thus reducing trial failures and speeding new treatments to market; and (3) Analyzing clinical trials and patient records to identify follow-on indications and discover adverse effects before products reach the market.

*c) Public Health*

(1) analyzing disease patterns and tracking disease outbreaks and transmission to improve public health surveillance and speed response; (2) faster development of more accurately targeted vaccines, e.g., choosing the annual influenza strains; and, (3) turning large amounts of data into actionable information that can be used to identify needs, provide services, and predict and prevent crises, especially for the benefit of populations.

*d) Evidence-based Medicine*

Combine and analyze a variety of structured and unstructured data-EMRs, financial and operational data, clinical data, and genomic data to match treatments with outcomes, predict patients at risk for disease or readmission and provide more efficient care.

*e) Genomic Analytics*

Execute gene sequencing more efficiently and cost effectively and make genomic analysis a part of the regular medical care decision process and the growing patient medical record. Analysis of DNA sequences is important in preventing evolution of viruses, bacteria during an early phase. Extract information from a large dataset causes the biggest challenge. And there is loss of information while analyzing a large sequence. Time consumption is high while reading a huge sequence. By using this physicians can predict the disease, thus it is useful for physicians and also prevent patients from developing the disease at the early stages. To overcome the above issues, the author's idea is to predict the disease through DNA sequence by initially visualizing the sequence into a graphical format and identify the normal structure and disease affected structure. Then to convert them into numerical value and perform various Machine learning algorithm (KNN, Linear regression, SVM, Logical regression) to find out the efficient algorithm with better accuracy. And finally to implement that with hadoop and predict disease.

*f) Pre-adjudication Fraud Analysis*

Rapidly analyze large numbers of claim requests to reduce fraud, waste and abuse.

*g) Device/Remote Monitoring*

Capture and analyze in real-time large volumes of fast-moving data from in-hospital and in-home devices, for safety monitoring and adverse event prediction.

Patient profile analytics: Apply advanced analytics to patient profiles (e.g., segmentation and predictive modeling) to identify individuals who would benefit from proactive care or lifestyle changes, for example, those patients at risk of developing a specific disease (e.g., diabetes) who would benefit from preventive care.

## 5.23  Future Scope

The future scope of this book is as follows: Liver images in this research could be extended for considering other category of diseases. More level of issues arises in the preprocessing part that could also be addressed as part of the research. Following the current trend, it is possible to apply Image Fusion for liver images to improve the representation of information. Multimodal Image Registration could also be analyzed and proposed in the work. Feature Extraction could be done with other techniques. Feature Reduction can be applied with Feature Extraction using suitable algorithms. Detailed analysis can be done on the nature of information in liver images and other unsupervised techniques could be suggested. Image Retrieval could also be applied along with classification and more importance can be given to storage and retrieval. Anthropometric data could also be considered during research. Fuzzy classification can be done with other membership functions. Fuzzy classifier can be modified with Neuro Fuzzy classifier and their performance could be analyzed. A complete integration of all the steps in the book could be provided as a single system to the doctor community.

## 5.24 MATLAB

MATLAB (matrix laboratory) is a numerical computing environment and fourth-generation programming language. Developed by MathWorks, MATLAB allows matrix manipulations, plotting of functions and data, implementation of algorithms, creation of user interfaces, and interfacing with programs written in other languages, including C, C++, Java, and FORTRAN. Although MATLAB is intended primarily for numerical computing, an optional toolbox uses the MuPAD symbolic engine, allowing access to symbolic computing capabilities. An additional package, Simulink, adds graphical multi-domain simulation and Model-Based Design for dynamic and embedded systems. In 2004, MATLAB had around one million users across industry and academia. MATLAB users come from various backgrounds of engineering, science, and economics. MATLAB is widely used in academic and research institutions as well as in industrial enterprises.

MATLAB stands for Matrix Laboratory. According to The MathWorks, its producer, it is a "technical computing environment". We will take the more mundane view that it is a programming language. This section covers much of the language, but by no means all. We aspire at the least to promote a reasonable proficiency in reading procedures that we will write in the language but choose to address this material to those who wish to use our procedures and write their own programs.

GUIDE (GUI development environment) provides tools for designing and programming GUIs. Using the GUIDE Layout Editor, you can graphically design your GUI. GUIDE then automatically generates the MATLAB code that defines all component properties and establishes a framework for GUI callbacks (routines that execute when a user interacts with a GUI component).

Matlab is a script that runs the main MATLAB executable on Microsoft Windows platforms. (In this chapter, the term Matlab refers to the script and MATLAB refers to the main executable.) Before actually initiating the execution of MATLAB, it configures the run-time environment by:

- Determining the MATLAB root directory
- Determining the host machine architecture
- Selectively processing command line options with the rest passed to MATLAB
- Setting certain MATLAB environment variables

There are two ways in which you can control the way MATLAB works:

- By specifying command line options
- By setting environment variables before calling the program

# References

Abd-Elmoniem, K.Z., Youssef, A.M. and Kadah, Y.M. 2002. Real-time speckle reduction and coherence enhancement in ultrasound imaging via nonlinear anisotropic diffusion. IEEE Transaction on Biomedical Engineering 49(9): 997–1014.

Aczel, J. and Daroczy, Z. 1975. On measures of information and their characterizations. New York: Academic.

Ahmadian, A. 2003. An efficient texture classification algorithm using gabor wavelet. IEEE Xplore Conference Engineering in Medicine and Biology Society, 2003. Proceedings of the 25th Annual International Conference of the IEEE, Volume: 1.

Aiazzi, B., Alparone, L., Baronti, S. and Lotti, F. 1998. Multiresolution local statistics speckle filtering based on a ratio laplacian pyramid. IEEE Trans. Geosci. Remote. Sens. 36(5): 1466–1476.

Albregtsen, F. 2008. Statistical Texture measures computed from Gray level co-occurrence matrices. Albregtsen: Texture Measures Computed from GLCM-Matrices, pp. 1–14.

Amores, J. and Radeva, P. 2005. Medical image retrieval based on plaque appearance and image registration. Plaque imaging: Pixel to molecular level, Suri, J.S. (Eds.). IOS press, pp. 26–54.

Aschkenasy, V. Schlomo, Christian Jansen, Remo Osterwalder, Andre Linka, Michael Unser, Stephan Marsch and Patrick Hunziker. 2006. Unsupervised image classification of medical ultrasound data by multiresolution elastic registration. Ultrasound in Med. & Biology 32(7): 1047–1054.

Aube, C., Oberi, F., Korali, N., Namour, M.A. and Loisel, D. 2002. Seminars in Ultrasound, CT, and MRI 23: 3.

Bleck, J.S., Ranft, U., Gebel, M., Hecker, H., Westhoff-Bleck, M., Thiesemann, C., Wagner, S. and Manns, M. 1996. Random field models in the textural analysis of ultrasonic images of the liver. IEEE Transactions on Medical Imaging 15(6): 796–801.

Breiman, L., Friedman, J., Olshen, R. and Stone, P. 1992. Classification and Regression Trees. Belmont. CA: Wadsworth International Group.

Brodley, C.E. and Utgoff, P.E. Multivariate Decision Trees. 19(1): 45–77.

Brown Gottesfeld, L. 1992. Survey of image registration techniques. ACM Computing Surveys 24(4): 325–376.

Brown, J.J., Naylar, M.J. and Yagan, N. 1997. Imaging of hepatic cirrhosis. Radiology 202(1): 1–16.

Burckhardt, C. 1978. Speckle in ultrasound B-Mode scans 25(1): 1–6.

Burt, P.J. and Adelson, E.A. 1983. The laplacian pyramid as a compact image code. IEEE Transactions on Communications 31(4): 532–540.

Carrillo, A., Cleveland, O.H., Duerk, J.L., Lewin, J.S. and Wilson, D.L. 2000. Semiautomatic 3-D image registration as applied to interventional MRI liver cancer treatment. IEEE Transactions on Medical Imaging 19(3): 175–185.

Catte, F., Lions, P.L., Morel, J.M. and Coll, T. 1992. Image selective smoothing and edge detection by nonlinear diffusion. SIAM J. Numer. Anal. 29(1): 182–193.

Chang, E. and Beitao Li, B. 2003. MEGA—The Maximizing Expected Generalization Algorithm for Learning Complex Query Concepts. ACM Transactions on Information Systems 21(4): 347–382.

Chen, E.L., Chung, P.-C., Chen, C.-L., Tsai, H.-M. and Chang, C.-I. 1998. An automatic diagnostic system for CT liver image classification. IEEE Transactions Biomedical Engineering 45(6): 783–794.

Chen, S.-T., Guo, Y.-J., Huang, H.-N., Kung, W.-M., Tseng, K.-K. and Tu, S.-Y. 2014. Hiding patients confidential data in the ECG signal via transform-domain quantization scheme. Journal of Medical Systems, June 2014, 38(6): 1–8.

Chen, Y., Yin, R.M., Flynn, P. and Broschat, S. 2003. Aggressive region growing for speckle reduction in ultrasound images. Pattern Recognition Letter 24(4-5): 677–691.

Chien-Cheng Lee, Yu-Chun Chiang, Chun Li Tsai and Sz-Han chen. 2007. Distinction of liver diseases from CT images using Kernel based classifiers. IC-MED 1(2): 113–120.

Chien-Cheng Lee, Sz-Han Chen and Yu-Chun Chiang. 2007. Classification of liver disease from CT images using a support vector machine. Journal of Advance Computational Intelligence and Intelligent Informatics 11(4): 396–402.

Collignon, A., Maes, F., Delaere, D., Vandermeulen, D. et al. 1995. Automated multimodality image registration using information theory. International Conference on Information Processing in Medical Imaging (IPMI, 95), IIe de Berder, France.

Cox, J. Kilian, Leighton, T. and Shamoon, T. 1997. Secure spread spectrum watermarking for multi-media. IEEE Trans. on Image Processing 6: 1673–1687.

Czerwinski, R.N., Jones, D.L. and Brien, W.D. 1995. Ultrasound speckle reduction by directional median filtering. International Conference on Image Processing 1: 358–361.

Deserno, T., Antani, S. and Long, R. 2008. Ontology of gaps in content based medical image retrieval. Journal of Digital Imaging.

Duda, R.O., Hart, P.E. and Stork, D.G. 2001. Pattern Classification, 2nd Edition. John Wiley & Sons, New York.

Dutt, V. and Greenleaf, J.F. 1996. Adaptive speckle reduction filter for log-compressed B-Scan images. IEEE Transactions on Medical Imaging 15(6): 802–813.

Edward Jero, S., Palaniappan Ramu and Ramakrishnan Swaminathan. 2014. Discrete wavelet transform and singular value decomposition based ECG steganography for secured patient information transmission. J. Med. Syst. 38: 132.

Edward Jero, S., Palaniappan Ramu and Ramakrishnan Swaminathan. 2015. ECG steganography using curvelet transform. Biomedical Signal Processing and Control Journal, September 2015, 22: 161–169.

Edward Jero, S. and Palaniappan Ramu. 2016. A Robust ECG steganography method. International Symposium on Medical Information and Communication Technology (ISMICT), March 2016.

Edward Jero, S., Palaniappan, Ramu and Ramakrishnan Swaminathan. 2016. Imperceptibility—Robustness tradeoff studies for ECG steganography using Continuous Ant Colony Optimization. Journal of Expert Systems with Applications, May 2016, 49: 123–135.

Egecioglu, O., Ferhatosmanoglu, H. and Ogras, U. 2004. Dimensionality reduction and similarity computation by inner-product approximations. IEEE Transactions on Knowledge and Data Engineering 16(6): 714–726.

E-Liang Chen, Pau-Choo Chung, Ching-Liang Chen, Hong-Ming Tsai and Chein I. Chang. 1998. An automatic diagnostic system for CT liver image classification. IEEE Transactions Biomedical Engineering 45(6): 783–794.

Fookes, C. and Bennamoun, M. 2002. The use of mutual information for rigid medical image registration: A review. In Proceedings of the IEEE Conference on Systems, Man and Cybernatics 4: 689–694.

Foster, K.J., Dewbury, K.C., Griffith, A.H. and Wright, R. 1980. The Accuracy of ultrasound in the detection of fatty infiltration of the liver. British Journal of Radiology 53: 440–442.

Francisco, P.M., Oliveira and Joao Manuel R.S Tavares. 2012. Medical image registration: a review. Computer methods in biomechanics and biomedical engineering DOI: 10.1080/10255842.2012.670855 (in press).

Frederik, M., Vandermeulen, D. and Suetens, P. 2003. Medical image registration using mutual information. Proceedings of the IEEE 91(10): 1699–1722.

Galloway, R.L., McDermott, B.A. and Thurstone, F.L. 1988. A frequency diversity process for speckle reduction in real-time ultrasonic images. IEEE Transaction on Ultrasonics 35: 45–49.

Geert Litjens, Thijs Kooi, Babak Ehteshami Bejnordi, Arnaud Arindra Adiyoso Setio, Francesco Ciompi, Mohsen Ghafoorian, Jeroen A.W.M. van der Laak, Bram van Ginneken and Clara I. Sanchez. 2017. A survey on deep learning in medical image analysis. arXiv preprint arXiv: 1702.05747.

Gletsos, M., Stavroula, G., Mougiakakou, George, K., Matsopoulos, Konstantina, S., Nikita, Alexandra, S., Nikita and Dimitrios Kelekis. 2003. A computer-aided diagnostic system to characterize CT focal liver eesions: design and optimization of a neural network classifier. IEEE Transactions on Information Technology in Biomedicine 7(3): 153–162.

Goshtasby, A. 2005. 2D and 3-D Image Registration for Medical Remote Sensing and Industrial Applicatqions, First Edition. Hoboken, NJ: J. Wiley & Sons.

Guild, M.O., Thies, C., Fischer, B. and Lehmann, T.M. 2007. A generic concept for the implementation of medical image retrieval systems. International Journal of Medical Informatics, pp. 252–259.

Guld, O.M., Thies, C., Fischer, B. and Lehmann, T.M. 2007. A generic concept for the implementation of medical image retrieval systems. Journal of Medical Informatics 76: 252–259.

Guodong Feng, Hongyan Shan, Yanqun Fei, Yanfu Huan and Qiang Fei. 2014. Quantitative near-infrared spectroscopic analysis of trimethoprim by artificial neural networks combined with modified genetic algorithm in Chemical Research in Chinese Universities, Springer 30(4): 582–586.

Gupta, N., Swamy, M.N.S. and Ploltkin, E. 2005. Despeckling of medical ultrasound images using data and rate adaptive lossy compression. IEEE Transactions on Medical Imaging 24(6): 743–754.

Hajduk, V., Broda, M., Kova, O. and Levicky, D. 2016. Image steganography with using QR code and cryptography. 26th Conference Radioelektronika, April 2016, pp. 350–353.

Hajnal, J.V., Hill, D. and Hawkes, D.J. 2001. Medical Image Registration, First Edition. New York: CRC Press.

Hall, M.A. and Smith, L.A. 1998. Practical feature subset selection for machine learning. In Proceedings of the 21st Australian Computer Science Conference, pp. 181–191.

Han, J., Kamber, M. and Pei, J. 2011. Data Mining: Concepts and Techniques, the Morgan Kaufmann Series in Data Management Systems, Morgan Kaufmann Publishers.

Haralick, R. 1979. Statistical and structural approaches to texture. Proceedings IEEE 67: 786–804.

Haralick, R.M., Shanmugam, K. and Dinstein, I. 1973. Textural features for image classification. IEEE Transactions on Systems, Man and Cybernatics 3(6): 610–621.

Harbin, W.P., Robert, N.J. and Ferrucci, J.T. 1980. Diagnosis of cirrhosis based on regional changes in hepatic morphology: a radiological and pathological analysis. Radiology 135(2): 273–83.

Hastie, T., Tibshirani, R. and Friedman, J. 2001. Elements of Statistical learning: Datamining, Inference, and Prediction. Springer Series in Statistics.

Herrick, J.F. 1959. Medical ultrasonics: Applications of ultrasound to biologic measurements. IRE Proceedings 47(11): 1967–1970.

Hill, D.L.G., Studholme, C. and Hawkes, D.J. 1994. Voxel similarity measures for automated image registration. pp. 205–216. *In*: R.A. Robb (Ed.). Proc. SPIE Visualization Biomedical Computing vol. 2359. Rochester, MN.

Hill, D.L.G., Studholme, C. and Hawkes, D.J. 1999. An overlap invariant entropy measure of 3D medical image alignment. Pattern Recognition 32(1): 71–86.

Hong, M.H., Sun, Y.N. and Lin, Z. 2002. Texture feature coding method for classification of liver sonography. Comp. Med. Imaging. Graph 6: 33–42.

Hong, P., Tian, Q. and Huang, T.S. 2000. Incorporate support vector machines to content-based image retrieval with relevant feedback. In Proc., IEEE Inter Conf. on image processing (ICIP'00) Vancouver, BC, Canada.

Hooke, R. and Jeeves, T.A. 1961. A direct search solution of numerical and statistical problems. J. Assoc. Comput. Mach, pp. 212–229.

Horng, M.H., Sun, Y.N. and Lin, X.Z. 2002. Texture feature coding method for classification of liver sonography. Comput. Med. Imag. Graph. 26: 33–42.

Hsu, C.W. and Lin, C.J. 2002. A comparison of methods for multiclass support vector machines. IEEE Transactions on Neural Networks 13(2): 415–425.

Huang, T.S. and Zhou, X.S. 2001. Image retrieval by relevance feedback: from heuristic weight adjustment to optimal learning methods. In Proc., IEEE Intern Conf on image processing (ICIP'01), Thessaloniki, Greece.

Ibaida, A. and Khalil, I. 2013. Wavelet-based ECG steganography for protecting patient confidential information in point-of-care systems. IEEE Transactions on Biomedical Engineering, December 2013, 60(12): 3322–3330.

Inderpal Bhander and Kevin Normandeau. 2013. Beyond Volume, Variety and Velocity are the issue of Big Data Veracity in Big_Data_Innovation_summit_Boston.

Jespersen, S., Wilhjelm, J. and Sillesen, H. 2000. *In vitro* spatial compound scanning for improved visualization of arthrosclerosis. Ultrasound Med. Biol. 26(8): 1357–1362.

Jesus Chamorro Martinez, Pedro Manuel Martinez Jimenez and Jose Manuel Soto Hidalgo. 2010. Retrieving Texture Images Using Coarseness Fuzzy Partitions. Hullermeier, E., Kruso, R. and Hoffmann, F. (Eds.): IPMU 2010, Part II, CCIS 81: 542–551.

Ji, Q., Engel, J. and Craine, E. 2000. Texture analysis for classification of Cervix lesions. Medical imaging. IEEE Transactions on 19(11): 1144–1149.

Jin Zhao and Bimal K. Bose. 2002. Evaluation of membership functions for fuzzy logic controlled induction motor drive. IEEE Xplore, IECON 02 [Industrial Electronics Society, IEEE 2002 28th Annual Conference] 1: 229–234.

John C. Platt. 1998. Sequential Minimal Optimization: A Fast Algorithm for Training Support Vector Machines. Technical Report MSR-TR-98-14, April 21, 1998.

Josien, P.W., Pluim, J.B., Antoine Maintz, Max A. and Viergever. 2003. Mutual information based registration of medical images: a survey. IEEE Transactions on Medical Imaging 22(8): 986–1004.

Kaplan, L. and Kuo, C.C. 1995. Texture roughness analysis and synthesis via extended self similar (ESS) model. IEEE Transaction on Pattern Analysis and Machine Intelligence 17(11): 1043–1056.

Kinkel, K., Lu, Y., Both, M., Warren, R.S. and Thoeni, R.F. 2002. Detection of hepatic metastases from cancers of the gastroinstestinal tract by using noninvasive imaging methods (US, CT, MR imaging, PET): a meta-analysis. Radiology 224: 748–56.

Khalifa Djemal. 2005. Speckle reduction in ultrasound images by minimization of total variation. IEEE International Conference on Image Processing 3(III): 357–360.

Klein Gebbinck, M.S., Verhoeven, J.T.M., Thijssen, J.M. and Schouten, T.E. 1993. Application of neural networks for the classification of diffuse liver disease by quantitative echography. Ultrasound Imag. 15: 205–217.

Kohonen, T. 1990. The Self-Organizing Map. Proceedings of the IEEE 78(9): 1464–1480.

Kuan, D.T., Sawchuk, A.A., Strand, T.C. and Chavel, P. 1985. Adaptive noise smoothing filter for images with signal dependent noise. IEEE Trans. Pattern Anal. Mach. Intell. 7(2): 165–177.

Kuncheva Ludmila, I. and Fried rich steimann. 1999. Fuzzy diagnosis. Artificial Intelligence in Medicine 16: 121–128.

Kwang In Kim, Keechul Jung, Se Hyun Park and Hang Joon Kim. 2002. Support vector machines for texture classification. IEEE Trans. Pattern Analysis and Machine Intelligence 24(11): 1542–1550.

Lagarias, J., Reeds, J.A., Wright, M.H. and Wright, P.E. 1998. Convergence properties of the Nelder-Mead simplex method in low dimensions. Society for Industrial and Applied Mathematics (SIAM J. OPTIM) 9(1): 112–147.

Laws, Kenneth Ivan. 1980. Textured image segmentation. No. USCIPI-940. University of Southern California Los Angeles Image Processing INST.

Lee, W. and Lee, C. 2008. A cryptographic key management solution for HIPAA privacy/security regulations. IEEE Trans. Inf. Technol. Biomed., Jan. 2008, 12(1): 34–41.

Lee, Yuan-Chang Chen and Kai-Sheng Hsieh. 2002. Ultrasonic liver tissues classification by fractal feature vector based on M-band wavelet transform. IEEE Transactions on Medical Imaging 22(3): 382–292.

Lehmann, T.M., Guld, G.O., Deselaes, T., Keysers, D., Schubert, H., Spitzer, K., Ney, H. and Wein, B.B. 2005. Automatic categorization of medical images for content based retrieval and data mining. Computerized Medical Imaging and Graphics 29: 143–155.

Little, R.J.A. and Rubin, D.B. 2002. Statistical analysis with missing data. Wiley Series in Probability and Statistics.

Maes, F., Collignon, D., Vandermeulen, D., Marchal, G. and Suetens, P. 1997. Multimodality image registration by maximization of mutual information. IEEE Transactions on Medical Imaging 16(2): 187–198.

Maintz, J.B.A., Meijering, E.H.W. and Viergever, M.A. 1998. General multimodal elastic registration based on mutual information. pp. 144–154. *In*: Hanson, K.M. (ed.). Medical Imaging: Image Processing. SPIE Press Bellingham, WA, vol. 3338 of Proc. SPIE.

McKinsey Global Institute (MGI). 2013. Big Data the next Frontier for Innovation, Competition, and Productivity.

Medina, J.M., Castillo, S.J., Barranco, C.D. and Campana, J.R. 2012. On the use of a fuzzy object relational database for flexible retrieval of medical images. IEEE Transactions on Fuzzy Systems 20(4): 786–803.

Metin Kahraman and Sevket Gumustekin. 2006. Mathching Aerial coastline images with map data using dynamic programming, 14th European Signal Processing Conference (EUSIPCO 2006), Florence, Italy.

Meyer, C.R., Boes, J.L., Kim, B., Bland, P.H., Zasadny, K.R., Kison, P.V., Koral, K., Frey, K.A. and Wahl, R.L. 1997. Demonstration of accuracy and clinical versatility of mutual information for automatic multimodality image fusion using affine and thin-plate spline warped geometric deformations. Medical Image Analysis I(3): 195–206.

Mohanty, S.P. 1999. Digital Watermarking : A Tutorial Review.

Momenan, R., Loew, M.H., Insana, M.F., Wagner, R.F. and Garra, B.S. 1990. Application of pattern recognition techniques in ultrasound tissue characterization. In Proc. 10th Int. Conf. Pattern Recognition 1: 608–612.

Mona Sharma. 2001. Evaluation of texture methods for image analysis. Intelligent Information System Conference. The Seventh Australian and New Zealand 2001.

Mousavi, S.M., Naghsh, A. and Abu-Bakar, S.A.R. 2014. Watermarking Techniques used in Medical Images: a Survey. Journal of Digital Imaging, December 2014, 27(6): 714–729.

Mukherjee, S., Chakravorty, A., Ghosh, K., Roy, M., Adhikari, A. and Mazumdar, S. 2007. Advanced Computing and Communications, ADCOM 2007 International Conference on; 2007. Corroborating the subjective classification of ultrasound images of normal and fatty human livers by the radiologist through texture analysis and SOM, pp. 197–202.

Muller, H., Michoux, N., Bandon, D. and Geissbuhler, A. 2004. A review of content based image retrieval systems in medical applications—clinical benefits and future directions. International Journal on Medical Information 73: 1–23.

Nadler, M. and Smith, E.P. 2001. Pattern Recognition Engineering. New York: Wiley.

Nicolas G. Rognin, Marcel arditi, Laurent mercier, Peter J.A. Frinking, Michel Schneider, Genevive Perrenoud, anass anaye, Jean-yves meuwly and Franois Tranquart. 2010. Parametric imaging of dynamic vascular patterns of focal liver lesions in contrast-enhanced ultra-sound. IEEE Transactions on Ultrasonics, Ferro Electricas, and Frequency Control, vol. 57, no. 11.

Ogawa, K., Fukushima, M., Subota, K. and Hisa, N. 2003. IEEE Trans. Nucl. Sci. 23: 3.

Oosterveld, B.J., Thijssen, J.M., Hartman, P.C. and Rosenbusch, G.J.E. 1999. Detection of diffuse liver disease by quantative echography: Dependence on a priori choice of parameters. Ultrasound Med. Biol. 19(1): 21–25.

Parkin, D.M., Bray, F., Ferlay, J. and Pisani, P. 2005. Global cancer statistics 2002. CA: A Cancer Journal for Clinicians 55(2): 74–108.

Pavlopoulos, S., Konnis, G., Kyiacou, E., Koutsouris, D., Zoumpoulis, P. and Theotokas, I. 1996. Evaluation of texture analysis techniques for quantitative characterization of ultrasonic images. *In*: 18th Annual International Conference of the IEEE Engineering in Medicine and Biology Society, pp. 1151–2.

Perona, P. and Malik, J. 1990. Scale-space and edge detection using anisotropic diffusion. IEEE Transactions on Pattern Anal. Mach. Intll. 12(7): 629–639.

Pluim, J.P.W., Maintz, J.B.A. and Viergever, M.A. 2000. Image registration by maximization of combined mutual information and gradient information. IEEE Transactions on Medical Imaging 19(8): 809–814.

Pluim, J.P.W., Antoine Maintz, J.B. and Viergever, M.A. 2003. Mutual-information-based registration of medical images: a survey. IEEE Transactions on Medical Imaging 22: 986–1004.

Pluim, J.P.W., Antoine Maintz, J.P. and Viergever, M.A. 2003. Mutual information based registration of medical images: a survey. IEEE Transactions on Medical Imaging 22(8): 986–1004.

Rolland, J.P., Muller, K. and Helvig, C.S. 1995. Visual search in medical images: A new methodology to quantify saliency. Proc. SPIE 2436, pp. 40–48.

Sattar, F., Floreby, L., Salomonsson, G. and Lovstrom, B. 1997. Image enhancement based on a nonlinear multiscale method. IEEE Transactions on Image Processing 6(6): 888–895.

Shin-Min Chao, Yuan-Ze Univ, Jungli, Du-Ming Tsai, Wei-Yao Chiu and Wei-Chen Li. 2010. Anisotropic diffusion-based detail preserving smoothing for image restoration. 17th IEEE Inter. Confer. on Image Processing, pp. 4145–4148.

Singh, A.K., Dave, M. and Mohan, A. 2014. Wavelet based image watermarking: futuristic concepts in information security. Proc. Natl. Acad. Sci., India, Sect. A Phys. Sci., September 2014, 84(3): 345–359.

Spendley, W., Hext, G.R. and Himsworth, F.R. 1962. Sequential application of simplex designs in optimization and evolutionary operation. Technometrics 4: 441–461.

Surya Prasath, V.B. and Arindama Singh. Well-posed inhomogeneous nonlinear diffusion scheme for digital image denoising. Journal of Applied Mathematics vol. Article ID 763847, 14 pages, 1010. DOI: 10.1155/2010/763847.

Thodi, D.M. and Rodriguez, J.J. 2007. Expansion embedding techniques for reversible watermarking. IEEE Trans. Image Process 16(3): 721–730, February 2007.

Tian, J. 2002. Reversible Watermarking by Difference Expansion. Multimedia and Security Workshop at ACM Multimedia, December 6, 2002, Juan-les-Pins, France.

Togashi, J., Tamura, S., Sugawara, Y., Kaneko, J., Matsui, Y., Yamashiki, N., Kokudo, N. and Makuuchi, M. 2008. Recurrence of cholestatic liver disease after living donor liver transplantation. World J. Gastroenterol. 14: 5105–5109.

Tuceryan, M. and Jain, A.K. 1993. Texture analysis. Handbook of Pattern Recognition & Computer Vision, Chapter 2.1, World Scientific, Singapore.

Wagner, R., Smith, S., Sandrik, J. and Lopez, H. 1983. Statistics of speckle in ultrasound B-scans. 30(3): 156–163.

Wang, H., Zhang, W. and Yu, N. 2015. Protecting patient confidential information based on ECG reversible data hiding. J. Multimedia Tools and Applications, June 2015, pp. 1–15.

Warden, P. 2011. Big Data Glossary. O'Reilly, Printed in USA.

Wen-Li Lee, Yuan-Chang Chen and Kai-Sheng Hsieh. 2003. Ultrasonic liver tissues classification by fractal feature vector based on M-band wavelet transform. IEEE Transactions on Medical Imaging 22(3): 382–292.

Wen, P. 2008. Medical image registration based-on points, contour and curves. International Conference on Biomedical Engineering and Informatics, pp. 132–136.

William H. Sameer Antanib, L., Rodney Long and George R. Thomas. 2008. A web-based image retrieval system for large biomedical databases. International Journal of Medical Informatics.

Wyawahare, M.V., Patil, P.M. and Abhyankar, H.K. 2009. Image registration techniques: an overview. International Journal of Signal Processing, Image Processing, and Pattern Recognition 2(3): 11–28.

Xue, Z., Antani, S.K. and Long, L.R. 2005. Relevance feedback for spine X-ray retrieval. In proceedings of the 18th IEEE Symposium on Computer Based Medical System, pp. 197–202.

Yali Huang, Lanxun Wang and Caixia Li. 2010. Texture analysis of ultrasonic liver images based on wavelet transform and probabilistic neural network. IEEE International Conference on Bio Medical Engineering and Informatics, 2008.

Yasser, M., Kadah, A.A., Farag, J.M., Zurada, A.M., Badawi and Youssef, A. 1996. Classification algorithms for quantitative tissue characterization of diffuse liver disease from ultrasound images. IEEE Transactions on Medical Imaging 15(4): 466–478.

Yang, C.-Y. and Wang, W.-F. 2016. Effective electrocardiogram steganography based on coefficient alignment. J. Med. Syst. 40: 66.

Yang, P.M., Huang, G.T. and Lin, J.T. 1988. Ultrasonography in the diagnosis of benign diffuse parenchymal liver diseases: a prospective study. Taiwan Yi Xue Hui Za Zhi-J Formosan Med. Assoc. 87: 966–977.

Yeh, W.C., Haung, S.W. and Li, P.C. 2003. Ultrasound Med. Biol. 29: 1229.

Yongjian, Yu, Janelle, A., Molloy and Scott Acton, T. 2004. Generalized speckle reducing anisotropic diffusion for ultrasound imagery. 17th IEEE Symposium on Computer-based Medical Systems, pp. 279–284.

Yu, Y.J. and Acton, S.T. 2002. Speckle reducing anisotropic diffusion. IEEE Transaction on Image Processing 11(11): 1260–1270.

Yu, J., Tan, J. and Wang, Y. 2010. Ultrasound speckle reduction by a SUSAN-controlled anisotropic diffusion method. Pattern Recognition 43(9): 3083–3092.

Zhang, F., Yoo, Y.M., Zhang, L., Koh, L.M. and Kim, Y. 2006. Multiscale nonlinear diffusion and Shock filter for ultrasound image enhancement. IEEE Computer Society Conference on Computer Vision and Pattern Recognition 2: 1972–1977.

Zhang, F., Yoo, Y.M., Koh, L.M. and Kim, Y. 2007. Nonlinear diffusion in laplacian pyramid domain for ultrasonic speckle reduction. IEEE Transactions on Medical Imaging 26(2): 200–211.

Zhao, J. and Bose, B.K. 2002. Evaluation of membership functions for fuzzy logic controlled induction motor drive. IEEE Xplore, IECON 02 [Industrial Electronics Society, IEEE 2002 28th Annual Conference] 1: 229–234.

Zicari, R.V. 2012. Big Data: Challenges and Opportunities, Big Data Computing.

Zikopoulos, P.C., deRoos, D., Krishnan Parasuraman, Deutsch, T., Corrigan, D. and Giles, J. 2013. Harness the Power of Big Data, McGraw-Hill.

Zikopoulos, P.C., Eaton, C., deRoos, D., Deutsch, T. and Lapis, G. 2012. Understanding Big Data–Analytics for Enterprise Class Hadoop and Streaming Data, McGraw-Hill.

# Index

# Authors' Biography

Dr. R. Suganya is working as Assistant Professor in the Department of Information Technology, at Thiagarajar College of Engineering, Madurai. Before that she worked as lecturer in the Department of Computer Science and Engineering, P.S.N.A. College of Engineering & Technology, Dindigul from 2005–2006. She received a Bachelor of Engineering degree in Computer Science and Engineering from R.V.S. College of Engineering, Dindigul. She received a Master degree in Computer Science and Engineering from P.S.N.A. College of Engineering, Dindigul. She earned her Doctorate in 2014 from Anna University, Chennai. She has 12 years of teaching experience. Her areas of interest include Medical Image Processing, Big Data Analytics, Internet of Things, Theory of Computation, Compiler Design and Software Engineering. She has published in reputed and refereed International Journals and IEEE Conferences. She received Young women award in Engineering from Venus International Foundations and Best Familiar Faculty award from ASDF—South Indian ASDF Awards 2017.

Rajaram was born in Mamsapuram near Rajapalayam in the year 1973. He received a B.E. in ECE in 1994 from the Thiagarajar College of Engineering, Madurai and a Master's degree with Distinction in Microwave and Optical Engineering from Alagappa Chettiar College of Engineering and Technology, Karaikudi in 1996. Dr. S. Rajaram holds a Ph.D degree in VLSI Design from Madurai Kamaraj University. He completed his Post Doctoral Research in 3D wireless system at Georgia Institute of Technology, Atlanta, USA during 2010–2011. Since 1998, he has been with Thiagarajar College of Engineering, Madurai. Currently he holds the post of Associate Professor in the department of Electronics and Communication Engineering, Thiagarajar College of Engineering. He is a former Member of Academic Council of Thiagarajar College of Engineering and Member of Board of Studies for several educational Institutions. His fields of interest are VLSI Design and Wireless Communication. Under his guidance ten research scholars have already obtained PhD degrees.

A. Sheik Abdullah, working as Assistant Professor, Department of Information Technology, Thiagarajar College of Engineering, Madurai, Tamil Nadu, India. He completed his B.E. (Computer Science and Engineering), at Bharath Niketan Engineering College, and M.E. (Computer Science and Engineering) at Kongu Engineering College under Anna University, Chennai. He is pursuing his Ph.D in the domain of Medical Data Analytics at Anna University Chennai. His research interests include Medical Data Research, E-Governance and Big Data. He has been awarded as gold medalist for his excellence in the degree of Post Graduate, in the discipline of Computer Science and Engineering by Kongu Engineering College. He has handled various E-Governance government projects such as automation system for tracking community certificate, birth and death certificate, DRDA and income tax automation systems. He has received the Honorable chief minister award for excellence in E-Governance for the best project in E-Governance for the academic year 2015–16.

Printed and bound by CPI Group (UK) Ltd, Croydon, CR0 4YY

01/11/2024

01782623-0020